Managing innovation in construction

Martyn Jones and Mohammed Saad

Thomas Telford

Published by Thomas Telford Publishing, Thomas Telford Ltd,
1 Heron Quay, London E14 4JD.
URL: http://www.thomastelford.com

Distributors for Thomas Telford books are
USA: ASCE Press, 1801 Alexander Bell Drive, Reston, VA 20191-4400, USA
Japan: Maruzen Co. Ltd, Book Department, 3–10 Nihonbashi 2-chome, Chuo-ku, Tokyo 103
Australia: DA Books and Journals, 648 Whitehorse Road, Mitcham 3132, Victoria

First published 2003

Also available from Thomas Telford Books

Managing construction supply chains. Andrew Cox and Paul Ireland.
ISBN 0 7277 3001 0
Best practice partnering in social housing development. Martyn Jones and Vic O'Brien.
ISBN 0 7277 3219 6

A catalogue record for this book is available from the British Library

ISBN: 0 7277 3002 9

Typeset by Keyword Typesetting Services, Wallington, Surrey

Foreword

The construction industry is important for everyone. As the Authors point out at the start of this book, there is no one untouched by it. Its competence is thus of interest to us all who are concerned with the development of common prosperity. We should therefore be keen to see innovation managed well in the industry and also alarmed that at present in the UK, construction companies reportedly spend more each year on litigation than on research and development.

This is where Martyn Jones and Mohammed Saad have shown great timing: there was never a better time to urge managers to innovate in this industry. The book is more than a practical guide, however. Importantly for established academic researchers, the authors provide a rich and well balanced diet of theory and concepts, developing them, in conclusion, into a set of models that are both conceptually elegant and practically clear.

This is a book, then, for managers and students alike. In its discussion, one can learn a great deal about its conceptual bases – innovation, supply chain management, project management and learning – while building an understanding of the context from the many engaging brief cases from the practical world. Every author wants to achieve the right balance in this respect – to keep every reader's attention; Jones and Saad have done just that. It is especially pleasing to find an account of the historical issues that have led the industry to where it stands today.

Students will find the book a valuable compendium as well as a thought-provoking read. Managers should find it of immediate practical use, as well as, one hopes, an invitation to reflect on the principles that underpin good practice.

Professor R C Lamming
Centre for Research into Strategic Purchasing and Supply (CRiSPS)
University of Bath

Dedication

To our wives and children whose lives have so enriched our own.

Acknowledgements

This book would not have been possible without the generous encouragement and support of a number of people. A particular thanks to John Edge and Peter James for their energetic and enthusiastic support over many years. We wish to record our thanks and debt of gratitude for the help and support received from the following:

- The late Paul Townsend, former Director of the Reading Construction Forum who believed in and supported the development of our initial ideas.
- Colleagues in the Faculty of the Built Environment and the School of Operations and Information Management at Bristol Business School of the University of the West of England, Bristol.
- We are particularly grateful to Laurence Krantz and John Edge for their contributions to Chapter 5.
- Those practitioners who have kindly provided information from their own organisations and given permission to reproduce material in the book.
- The editorial and production teams at Thomas Telford Publishing, including James Murphy and Jeremy Brinton for their patience and encouragement.
- The reviewers, Professors Richard Lamming (Bath School of Management), Peter Hines (Cardiff Business School) and James Barlow (Science Policy Research Unit, Sussex University) and Peter Appleby (Spen Hill Properties/Tesco), for their constructive and perceptive comments on the original structure and content of the book.
- Last, but by no means least, our very special thanks go to our wives, children and families for their unstinting and unconditional love and support during the research and the writing of this book, most especially for the many weekends they had to spend without us.

We are all influenced by the thoughts and ideas of other people which tend to drift into the subconscious and are not always distinguished clearly from our own. We have attempted to reference the sources of work by other writers but apologise to any concerned if their contributions have not been adequately recorded. Despite all the help and assistance that we have received, any faults or shortcomings in the final product are ours and ours alone.

Contents

List of figures

List of tables

List of case studies

Introduction

This book is about gaining a better understanding of innovation, its key determinants and implementation process. Its dual objectives are to contribute to the general on-going debate on innovation and its management, and to investigate the promotion, development, support, implementation and sustaining of innovation in construction.

Innovation is a new idea that leads to enhanced performance. It is not a single nor an instantaneous act but a whole sequence of events that occurs over time and involves all the activities of bringing a new product, process or service to the market. It is regarded as a complex sequence of events involving many different functions, actors and variables and forming a process which is not reducible to simple factors. This complexity explains the adoption of a multidisciplinary approach by the authors in their ten years of research on innovation in construction. This multidisciplinary perspective emanates from a synergistic combination of the two disciplines of management and construction.

In spite of the introduction of many successful innovations, much of the UK construction industry is still seen as being slow to innovate. At its best, UK construction is excellent, but it is still underperforming in meeting its own needs and those of its clients. This relatively poor performance of construction can be attributed to a number of features that set it apart from other industries. Although construction embodies a complex mix of activities, relationships, knowledge and skills, the industry is overstating its uniqueness and is underperforming in comparison with many other industries. A number of reports have identified the strengths and weaknesses in the UK construction industry. Most of these reports have highlighted essentially the same persistent weaknesses.

Changes in construction have been shaped by the emergence of Fordism and post-Fordism paradigms. If the central characteristic of Fordism was the mass production of standardised products, post-Fordism is characterised by:

- increasingly flexible labour processes and markets
- heightened geographical mobility
- rapid shifts in patterns of consumption
- privatisation, deregulation and a reduction in state intervention
- the greater use of programmable Information, Communication Technology (ICT)
- more integrated and holistic approaches
- more responsiveness to the needs of customers
- a greater recognition of knowledge as a competitive advantage
- a significant shift in the nature of competitive advantage from the physical to the intangible and from seen to the unseen.

There is evidence that clusters of interrelated technical, organisational and managerial innovations, based on Fordist principles, coupled with ideas imported from other industries, led to innovation in products and processes. However, there was also resistance to fully embrace this approach because of the industry's conservatism in relation to change and its defining characteristics.

With the emergence of the post-Fordism approach from the early 1980s, construction's products and processes have become more sophisticated and the pressures on the industry to improve have intensified. Parts of construction are responding to the new market pressures and the main principles associated with the new paradigm. There is some evidence that construction has improved its performance since the mid-1990s with an increase in the variety of products and services within certain sectors.

Despite considerable pressure for change from the wider environment and the regulatory and institutional framework, much of the industry is still continuing to be structured and constrained by the economic, social and technological trade-offs defined by Fordism. This would indicate that much of the industry remains untouched by post-Fordism. This can be attributed to a number of factors, but most significantly the lack of in-depth understanding of the complexity of innovation as an evolutionary, dynamic, lengthy and complex process. Construction's difficulties in effectively managing innovation can be explained by the limited level of understanding of a number of key factors including the need:

(*a*) to adopt a strategic and systematic approach in developing, implementing, monitoring and sustaining innovation
(*b*) to create and sustain an organisational culture for learning and innovation
(*c*) to ensure top management commitment and acceptance of risk
(*d*) to promote and diffuse innovation through key individuals or champions

(*e*) to set up good linkages within and between organisations leading to more collaborative relationships and
(*f*) to exploit external support.

Project management, Partnering and Supply Chain Management are examined in order to review how construction has implemented, managed and sustained innovation in making the transition from Fordism to post-Fordism.

There are a number of critical issues to be addressed by project management in responding to post-Fordism and the key features of innovation including the short-term and adversarial relationships, and the tensions and trade-offs between time, cost and performance that characterise projects. The current mind set in project management suggests that it cannot fully respond to the new paradigm in which all participants and stakeholders need to be fully involved. It also requires more collaborative relationships and open modes of communication. Implementing and sustaining innovation in projects requires new management approaches and business practices. The success of these new approaches depends on a radical change in attitudes in order to develop more collaborative customer-supplier relationships, less adversarial and less contractual arrangements to encourage inter-organisational networking, sharing and learning.

The adoption of partnering is construction's first significant attempt to innovate through addressing the industry's adversarial relationships and fragmented processes, and hence responding to post-Fordism. Encouraged by substantial improvements in performance from first and second generation partnering, some parts of construction have started moving towards the adoption of third generation partnering which corresponds to Supply Chain Management (SCM). Extending a comprehensive SCM approach to the whole of the construction market is difficult and challenging. The present complexity, fragmentation, interdependency and uncertainty which characterise the construction market and process will influence the way in which SCM and other innovations are adopted and implemented. One of the primary objectives of this book is to reflect on the main inhibitors to innovation in construction and areas to be addressed in order to better manage and diffuse innovation within the industry.

The structure of the book comprises four parts, as illustrated in the diagram on page xvii.

Part 1, which comprises Chapters 1, 2 and 3, defines the construction context within which innovation takes place. Chapter 1 examines the specificity and nature of construction and highlights its strengths and weaknesses in managing change. It also identifies the contextual factors which need to be taken into account in managing innovation in construc-

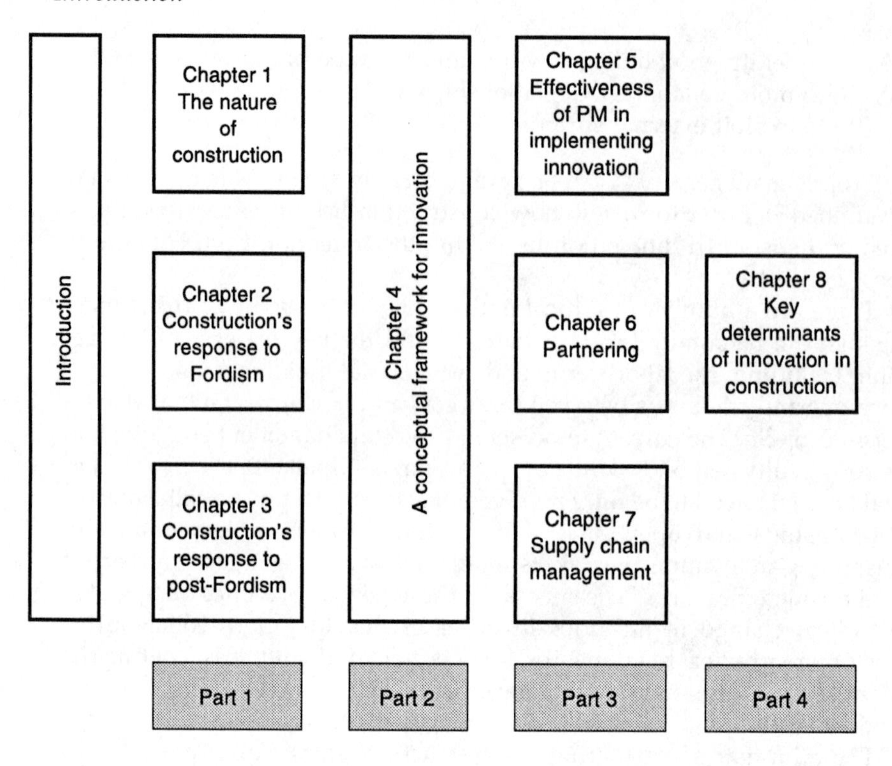

tion. Chapter 2 investigates, in the context of Fordism, how the key features of construction have shaped its ability to innovate and change. Chapter 3 analyses the emergence of post-Fordism and its impact on construction. It aims to review the way in which construction has introduced and managed innovation in its products and processes in order to respond to the challenges of the new paradigm.

Part 2, which consists of Chapter 4, aims to define a framework for innovation. It reviews the main theory and models of innovation to help understand the innovation process and its main determinants.

Part 3, which includes Chapters 5, 6 and 7, investigates Project Management, Partnering and Supply Chain Management in order to review how construction has implemented, managed and sustained innovation in making the transition from Fordism to post-Fordism, with reference to the theoretical framework of innovation and the nature of construction.

Chapter 5 examines the evolution of project management and how it has responded to Fordism and post-Fordism and questions its effectiveness as a tool for innovation. Chapter 6 is devoted to the study of partnering as a key innovation to improve project performance. It assesses the motives, relevance and effectiveness of partnering as an innovation.

Chapter 7 assesses the implementation and management of innovation in construction through the adoption of SCM.

Part 4, which comprises Chapter 8, brings together the key features of the theoretical framework for innovation and the context and specificity of construction. It reflects on the main determinants of innovation in construction and areas to be addressed in order to better manage and diffuse innovation within the industry.

Chapter 1

The key features shaping the context for innovation in construction

The construction industry comprises those enterprises and individuals whose main activity is in the construction and maintenance of the built environment.[1] The construction process comprises the set of diverse activities that are associated with the design, management, production, maintenance, refurbishment and eventual disposal of buildings and civil engineering works. This means that construction is a major economic sector in its own right and thus a key aspect of economic activity. As well as contributing to gross domestic product it also underpins all other sectors of the economy by providing the infrastructure and facilities essential for the rest of the economy and typically accounts for half of the UK's annual fixed capital formation. Its products and services are critical to the functioning of a modern economy and, if effective, it can provide an important economic multiplier. In turn, the prospects for the industry itself are highly dependent upon the success of the rest of the economy and government plans for public expenditure.

In the UK, as in most developed economies, the construction industry produces between 6 and 10% of GDP, and is the fourth largest in Europe, representing over 10% of total EU construction activity. When construction-related goods and services, such as materials components and equipment are included, the construction industry can represent more than 12% of the UK's GDP. The value of construction output in 2000 was approximately £58 billion with repair and maintenance of existing infrastructure and facilities accounting for nearly 50% of the total. In the same year it employed around 1.5 million people and also had a large number of employees involved indirectly, through work in materials and components manufacturing and in plant and vehicle building. Irrespective of

[1] This definition equates broadly with Category 42 of the Revised 1992 Standard Industrial Classification (SIC). However, it should be borne in mind that industries are notoriously difficult to define.

its size, the buildings and infrastructure provided by construction underpins other sectors of the economy and much of the social, political and economic life of advanced economies and plays a vital role in the competitiveness of most sectors of the UK economy. Indeed it is so important that it cannot just be the concern of those directly involved. Society as a whole has a direct and vested interest in its products, practices, efficiency and effectiveness and in ensuring that it is protected from its own excesses.

Like many other sectors of the economy, the industry has become increasingly global following the rapid growth in Middle East markets in the mid 1970s and the decline in domestic markets. This means that the largest practices and general and specialist contractors now operate internationally as well as nationally. These organisations have had to adapt to designing and constructing to international standards in diverse locations, in competition with practices and firms from other developed countries and using increasingly international supply systems.

Much of the UK construction industry, however, is still described as backward. A number of researchers have identified the industry's shortcomings over a considerable period of time. For example, 'technologically stagnant' (Business Roundtable 1982), 'fragmented' (Barrie and Paulson, 1978), and 'negligible R and D' (National Economic Development Office, 1985). It is perceived by its customers as being 'slow' and 'costly' (NEDO, 1985 and *Financial Times*, 7 January 1981) and delivering poor quality products and services in comparison with many other sectors of the economy (Cox and Townsend, 1998). Latham (1994), identified the 'high cost' of construction in the UK and called for a 30% reduction. Egan (1998), argued that although UK construction at its best is excellent, there is deep concern in the industry and among its clients that it is 'underachieving' both in meeting its own needs and those of its clients.

A succession of studies have identified the strengths and weaknesses of the industry. As can be seen from Table 1.1, there is considerable consensus of opinion on the strengths and weaknesses of construction. The strengths of the industry include its flexibility and adaptability. Its persistent weaknesses include its fragmented structure and project processes and its adversarial relationships. The emergence of post-Fordism and increased benchmarking of construction's performance against other industries has revealed further deficiencies in its performance in relation to customer focus, learning and leadership.

It has been argued that the relatively poor performance of construction can be attributed to a number of features that set it apart from other industries. For example, its product is bespoke, generally large, bulky and immobile. This means that it is largely produced at the point of consumption on premises owned by the client or customer. However,

Table 1.1. *Reports on the UK construction industry*

	Simon 1944	Emerson 1962	Tavistock 1963	Banwell 1964	Tavistock 1966	NEDO 1967	Latham 1994	Jones & Saad 1998	Egan 1998	Audit office 2001	Fairclough 2002	Accelerating change 2002
Technology												
Development of new materials												
Increased variety												
Increased technical complexity												
New processes												
Unwillingness to innovate												
Organisation												
Outstripped by technological developments												
Adherence to outmoded procedures												
Development of subcontracting/specialists												
Organisational independence												
Interdependence of activities												
Division of responsibility												
Relationships/cooperation												
Intercommunication												
Collaboration												

(continued)

Table 1.1. (contd.)

	Simon 1944	Emerson 1962	Tavistock 1963	Banwell 1964	Tavistock 1966	NEDO 1967	Latham 1994	Jones & Saad 1998	Egan 1998	Audit office 2001	Fairclough 2002	Accelerating change 2002
Integration of design and construction												
Partnering												
Uncertainty												
Feedback												
Flexibility												
Efficiency												
Low productivity and poor return												
Standardisation												
Construction as a manufacturing process												
Prefabrication and modularisation												
Pre-construction												
Role of the building owner												
Changing client base												
Public sector emphasis												
Private sector emphasis												
Fragmented client base												
Informed and naïve client distinction												

	Effective briefing												
	Effective planning and coordination												
	Management of design												
	Supplier selection												
	Criticism of competitive tendering – outmoded												
	Lack of final and detailed information												
	Questioning of value of bill of quantities												
	Reduction of subsequent changes												
	Selective tendering												
	Burdensome qualification procedures												
	Negotiation												
	Direct negotiation												
	Serial contracting												
	Site management												
	Lack of information												
	Training												
	Coordination role												
	Division of responsibility												

(continued)

Table 1.1. (contd.)

	Simon 1944	Emerson 1962	Tavistock 1963	Banwell 1964	Tavistock 1966	NEDO 1967	Latham 1994	Jones & Saad 1998	Egan 1998	Audit office 2001	Fairclough 2002	Accelerating change 2002
Benefit of early contractor input												
Reduction of changes												
Adoption of incentives												
Legal/contractual												
Increased litigation/ arbitration												
Increased claims and variations												
Prompt agreement – claims and variations												
Payment												
Prompt and regular payment												
Elimination of retention money												
Rethinking of fee structure												
Competitive fees												
Fees for specialist design												
Revamp of payment procedures												

	Organisational												
	Independent advice for clients												
	Conducive project environment												
	Leadership												
	Teamwork and collaboration												
	Learning												
	Prof. development for graduates into management												
	Qualified workforce												
	Empowerment												
	Use of Respect for People's toolkits												
	Innovation												
	Insurance to indemnify whole team												
	Supply chain management												
	Integrated processes and teams												
	Appropriate relationships												
	Customer focus												
	Value management												
	Use of process maps												
	Lean thinking												

(continued)

Table 1.1. (contd.)

		Simon 1944	Emerson 1962	Tavistock 1963	Banwell 1964	Tavistock 1966	NEDO 1967	Latham 1994	Jones & Saad 1998	Egan 1998	Audit office 2001	Fairclough 2002	Accelerating change 2002
	Innovation and research												
	Investment in R&D												
	Contribution to quality of life												
	Appropriate research												
	Excited researchers												
	Multi-disciplinary research teams												
	Interchange between industry and academia												
	Use of Teaching Company Scheme												
	Support best innovators												
	Environmental sustainability												
	Responsibility for sustainability of products												
	Education												
	Integrated team-working in courses												

there is a growing consensus that although construction embodies a complex mix of activities, relationships, knowledge and skills that are often not fully understood by outsiders or observers, the industry is overstating its uniqueness and is underperforming in comparison with many other industries. This relatively poor performance of construction cannot be attributed to a single, simple factor. Rather there are a number of weaknesses on both the demand and supply sides of the industry that have developed over many years. Demand-side weaknesses include the undue emphasis placed on price and time by clients, their poor approach to supplier selection and fluctuations in overall demand and between different sectors of the industry (Cox and Townsend, 1998). On the supply side, the Latham report contended that the industry's reputation for poor quality stems mainly from the low barriers to entry in general contracting and the low levels of investment in training.

The reluctance of the industry to change, in response to these reports (Latham), has given rise, in part, to the charge that the industry is 'backward' (Clarke, 1985), and is reluctant to innovate. A number of writers (Needleman, 1965 and Ball, 1988) suggest that 'the absolute physical constraint thesis' may be responsible for this reluctance to embrace change. According to this thesis, the advances associated with the manufacturing industry and other sectors of the economy are impossible to achieve in construction. For example, the uniqueness of each building and its production process, and the way breaks in the flow of production are forced on construction by the geographical spread of its projects present insurmountable barriers. This partly explains construction's diversity, uniqueness, complexity, fragmentation, and adversarial relationships.

Construction's large number of clients and their different requirements

On the demand side, the construction market is highly segmented in terms of the types of clients and the products and services they require. The most comprehensive and accessible source of information about the demand for construction products and services is provided by the Government Statistical Office. The *Construction Statistics Annual* published each year categorises construction work by type of client and type of work. Differentiating between public and private clients is becoming more problematic with the blurring of the boundaries between the two sectors with the privatisation of major utilities and other previously nationalised industries, the increased commercialisation of housing associations, and the increased use of the Private Finance Initiative (PFI). It is clear that public sector spending is a significant part of construction demand and that, while in 1989 the public sector was responsible for around one-quarter of new orders, by 1999 the share had reduced to just over one-fifth. The statistics also show the importance of maintenance and repair work, which represents approximately 50% of the total value of construction output. In the public sector, its share is around 60% of output and around 40% in the case of work in the private sector.

The categories of new work identified in the government statistics provide an indication of the diversity of construction's markets. These include:

- housing
- factories
- warehouses
- oil, steel and coal facilities
- schools and colleges
- universities, hospitals and clinics and other health facilities
- offices
- entertainment complexes
- garages
- agriculture buildings
- infrastructure and facilities for water, sewerage, electricity and communications
- railways
- harbours
- roads.

Figure 1.1 gives an indication of the proportions of work undertaken in each market segment for the year 2000. However, there is demand for an even wider variety of buildings and other structures than suggested by the broad categories shown in the national statistics.

The type of work in each market sector generates quite different businesses. In the case of private housing, housebuilders buy land (often building up large land banks) and developing and marketing housing to meet local market needs. They rarely undertake the construction work, often outsourcing it to smaller builders and subcontractors. The houses are mainly sold to owner-occupiers or to Housing Associations or Registered Social Landlords for social housing. In the case of property development, construction firms may work alone or with others. The land is purchased and an industrial or commercial building is designed and constructed and then let or sold. The new facility can be prelet, presold or developed in conjunction with the ultimate owner, who may be the occupier or an investor. Given the time taken to develop such properties, the risks can be substantial. In the case of contracting, construction companies do not normally invest

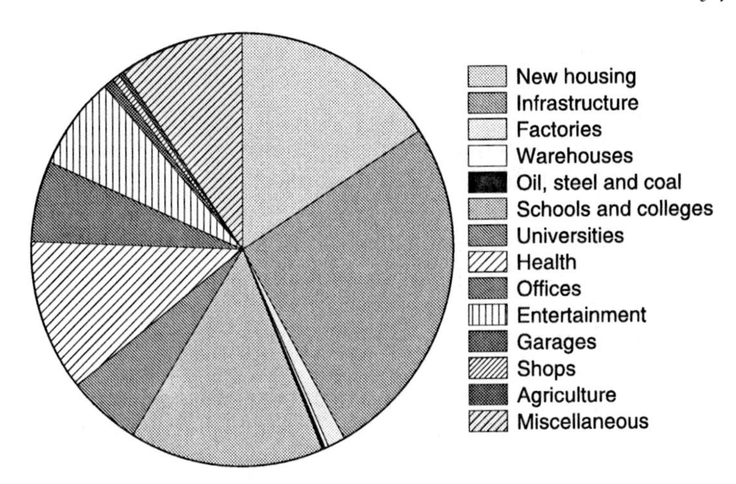

Fig. 1.1. Range and significance of the main market segments in construction

their own money in the project but undertake the work to the order of the client and risks can be lower than in property development and housebuilding. All these activities are influenced by the state of the economy, including the expectations of clients, rates of interest and the availability and cost of finance. They are also influenced by factors unrelated to the state of the economy such as changes in legislation or the planning regulations.

As well as generating types of businesses, each sector of the industry embodies a cluster of technologies and ways of working which differentiates it in some way from construction activities in other sectors. For example, housing essentially comprises concrete strip foundations, load-bearing masonry walls or timber frames, timber upper floors and pitched roof trusses supporting tiles. Whereas office buildings often comprise concrete pad foundations on which sits a steel or concrete structural frame to which some form of preformed or *in situ* wall cladding is attached.

As well as this variability between construction sectors and types of facilities, there are also differences in technologies, skills and ways of working between specific projects within each sector of the industry. The nature of these specific projects, the use for which they are intended, the dominant technologies and processes used in their construction are all major determinants of a specific project's supply chains and, collectively, the industry's overall supply systems. The characteristics of a specific project and hence its degree of uniqueness is determined by a number of factors including:

- its intended use, size, location, immediate and wider environment
- the client or project sponsor and the ultimate user

- the performance requirements and technological sophistication of the final product
- the legal and political environment
- volume of resource requirements and delivery requirements
- the spatial relationship between the project and the resource groupings
- the procurement strategies of the key participants including form of contract, commercial criteria, cost, quality and speed of delivery, allocation of risk and
- the timing of the resource inputs.

The assessment of these project characteristics determines the resources needed for a project and the selection of the most appropriate supply chains needed to deliver clusters of resources and services for the project as a whole.

This diversity and uniqueness means that construction projects are very often a one-off experience as the requirements and specifications of technologies for a specific client determine their characteristics. They involve assembling materials and components designed and produced by a multitude of suppliers, working in a diversity of disciplines and technologies, in order to create a facility or building for a specific client. Most buildings are constructed of many different materials and components which themselves are made of sub-systems. With the increasing shift from on-site to off-site production, managing construction projects involves integrating diverse and complex supply systems in which a growing amount of the value of the product is added. This diversity of product technologies, coupled with discontinuous demand from over 80% of construction's clients accounts for the transient nature of relationships between the demand and supply side of the industry.

In addition, many commercial clients have to respond flexibly and quickly to an increasingly competitive and rapidly changing business environment within the time-scale of even modest projects. This need to respond to rapidly changing business requirements often results in changes to the building whilst it is under construction, and explains why most of construction's procurement routes have been developed with this flexibility in mind.

The diversity of the construction and the variability in the production process has to be reorganised with each new construction project. For each project a series of decisions has to be taken which take into account the client's specific requirements and the context in which the tasks are to be executed. Temporary organisational structures have to be created, strategies developed and plans achieved. Many decisions are often of a technical nature, but some are more related to the logistics of procurement, the problems of communication, the application of scarce or expensive resources, legal and regulatory constraints, and the motivation of diverse groups of individuals and organisations. This can make project outcomes difficult to predict and can result in short and often adversarial relationships. The variability and segmentation of the construction market, coupled with the price-competitive approaches to procurement of most clients, means that the industry has to be more geared to flexibility, responsiveness and agility rather than ultimate product performance, which may well be more readily achieved in the longer-term and more stable processes associated with many other industries (Saad and Jones, 1995).

The diversity of clients, buildings, sites, materials and components means that the project processes have a wide range of supply systems supplying materials, components and sub-assemblies which account for nearly 60% by value of construction costs. It is the scope and nature of projects that determine the size and type of construction demand and the nature and magnitude of the supply systems to service them. Buildings normally comprise a number of separate elements such as substructure, superstructure, internal and external envelope, internal fixtures and fittings, mechanical and electrical services, and external works. Each of these elements or work packages has different functions and therefore comprises quite different technologies. The substructure supports building loads and transfers them to the soil on which the building sits. The activities associated with this work package include excavation, earthwork support, groundwater control, the casting of reinforced concrete and tanking. The main trades involved include ground working, steelfixers and concreting. The superstructure supports the external and internal envelope and then transfers live and dead loads to the substructure. The dominant technologies in this work package are steel and reinforced concreters in either precast or in-situ form. Activities include steel erection, falsework and formwork, steelfixing and concreting. The envelope provides shelter from the external environment, conserves energy, creates the appropriate shape and volume of internal space, and provides an acceptable internal and external appearance. Given the wide range of envelope technologies available, the activities and trades associated with this work package are diverse but include brick and block laying, panel erection, installing insulation, glazing, plastering and painting and decorating. The mechanical and electrical services are incorporated to enhance the internal environment through the provision of heating, lighting and mechanical ventilation, water supply and the disposal of waste as well as facilitating the movement of people and goods within the building. Trades include plumbers, electricians and heating and ventilation engineers.

Supply systems have developed to service the range of technologies and processes associated with the different parts of the building known as building elements or work packages. For each of these elements the most appropriate supply chains have to be selected, organised and aligned in a way that best fits with the client's requirements and the overall objectives and strategies for the project. As can be seen from Fig. 1.2, construction supply systems include construction organisations and supplier networks. The number of private contractors that were operating in construction in 2000 is shown in Table 1.2.

These supply systems provide incomplete sets of resources and services, parts of which are assembled to make up supply chains to individual elements or work packages. As well as being shaped by the technological requirements of each element, these supply chains are further conditioned and shaped by the complex linkages between individuals, firms and projects and the procurement strategies and forms of contract adopted by the main project participants, but particularly by the client. Projects and their supply chains are also influenced by wider institutional, economic, social, environmental and political factors.

There are major differences in the main participants, the managerial and technical skills, the plant and equipment and supply chains needed for each of the main elements or work packages. This results in major cost differentials between the various elements or work packages for a specific building and between the elements or work packages for different buildings. For example, the structure typically accounts for 15% of a building's cost but can vary from 5% for refurbishment works to 40% for multi-storey car parks. The internal and external envelope normally accounts for 25% of the total cost of a building but again is subject to

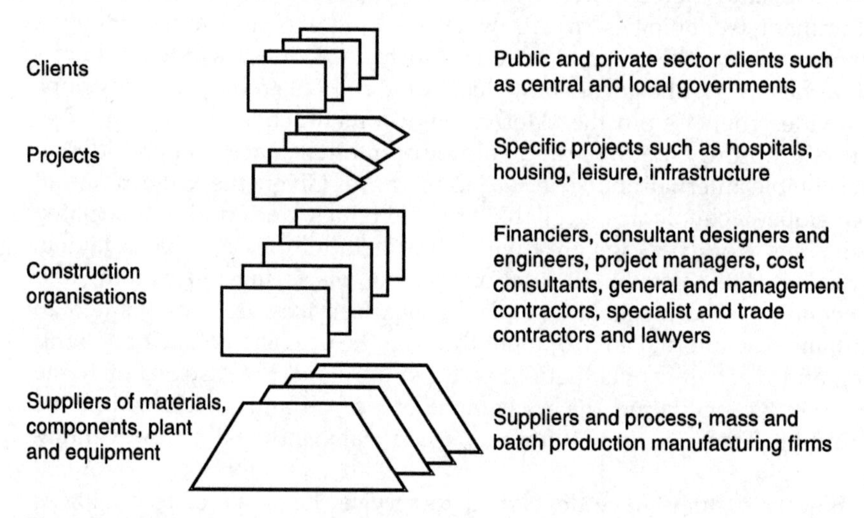

Fig. 1.2. Construction supply systems

Table 1.2. Types and numbers of construction firms in 2000

Type of firm	*Number of firms*
General trades	
General builders, building and civil engineering contractors, civil engineers	59 708
Specialist trades	
Constructional engineers	1105
Demolition	885
Reinforced concrete specialists	263
Roofing	6310
Asphalt and tar sprayers	845
Scaffolding	1555
Installation of electrical wiring and fitting	18 426
Insulating activities	879
Plumbing	19 973
Heating and ventilating engineers	5870
Plastering	2389
Joinery installation	9699
Flooring contractors	2820
Floor and wall tiling specialists	791
Suspended ceiling specialists	3452
Painting	8507
Glazing	3581
Plant hire (with operators)	3245
Other construction work and building installation and completion	19 154
All trades	163 426

variations between different buildings. For example, it can range from 19% in the case of hospitals and sports and leisure facilities to more than 30% for office developments, educational buildings and housing. Mechanical and electrical services typically account for 25% of total building costs but can exceed 50% for more highly serviced buildings (Drewer and Hazelhurst, 1998).

Fluctuating and discontinuous demand

Client demand dictates the nature and structure of the construction industry. Fluctuations in overall demand for construction products during the period 1990–2000 are shown in Fig. 1.3.

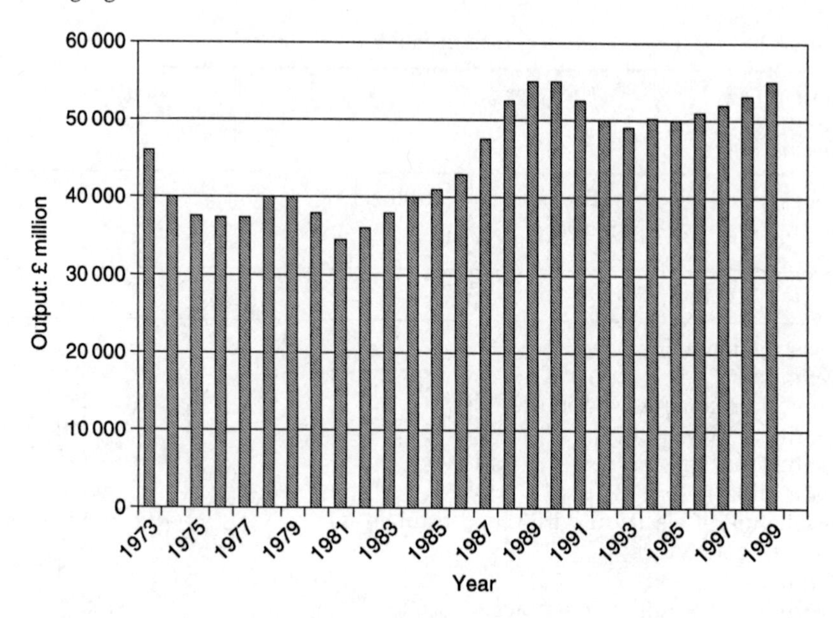

Fig. 1.3. Construction output 1980–1999 at 1990 constant prices

Figure 1.3 shows clear cycles of slump and relative boom. It has been argued that these fluctuations are greater than in many other sectors of the economy. There is no simple explanation for this phenomenon but it has been argued that successive governments, as major clients of the industry and regulators of the economy, have used the industry to pursue Keynsian policies, turning off investment in capital projects during periods of overheating in the overall economy and switching on investment to stimulate the economy during periods of recession. It also appears to arise because of shifts in levels of demand in the various sectors of the economy. Table 1.3 presents the output of construction, by sector, for the years 1983, 1990, 1993 and 2000. It shows how the output is divided into new and repair and maintenance (R&M) and the broad classifications of work undertaken by the industry.

As can be seen from Table 1.3, around 45% of construction output is in the R&M of existing facilities. It also shows that the proportion of construction output in each sector changes over time. For example, in 1983 public housing accounted for 4.3% of output, which declined to 1.7% in 1990, rose again to 3.3% in 1993 before falling back to 1.9% in 2000. Infrastructure new orders show a significant shift. In 1989, prior to a period of privatisation, the public sector was responsible for 80% of infrastructure spending. By 1999, this was down to 35%.

Table 1.3. Construction output by sector 1983, 1990, 1993 and 2000

1983 **Total output £38 980 million**		**1990** **Total output 55 308 million**	
Public housing	4.3%	Public housing	1.7%
Private housing	18.5%	Private housing	10.4%
Public infrastructure	7.1%	Public infrastructure	5.9%
Private infrastructure	0.2%	Private infrastructure	3.1%
Public non-housing	9.3%	Public non-housing	8.0%
Private industrial	4.8%	Private industrial	6.1%
Private commercial	10.6%	Private commercial	20.4%
Public and private housing R&M	24.2%	Public housing R&M	9.7%
		Private housing R&M	15.3%
Public non-housing R&M	13.7%	Public non-housing R&M	9.9%
Private non-housing R&M	7.2%	Private non-housing R&M	9.0%
1993 **Total output £48 502 million**		**2000** **Total output £69 527 million**	
Public housing	3.3%	Public housing	1.9%
Private housing	11.3%	Private housing	12.4%
Public infrastructure	8.0%	Public and private infrastructure	9.2%
Private infrastructure	6.0%		
Public non-housing	10.8%	Public non-housing	7.0%
Private industrial	5.4%	Private industrial	5.3%
Private commercial	13.5%	Private commercial	18.1%
Public housing R&M	9.8%	Public housing R&M	9.4%
Private housing R&M	14%	Private housing R&M	14.9%
Public non-housing R&M	8.8%	Public non-housing R&M	8.2%
Private non-housing R&M	9.0%	Private non-housing R&M	13.6%

The figures for contractors' output by type of work demonstrate the variability of workload in different sectors. Figure 1.4 shows the contractors' output for private offices which is particularly prone to fluctuations because of its close dependence on the general state of the economy.

There is a wide variation in the size and complexity of projects across the sectors. The demand side of the industry is characterised by a large number of small and technically unsophisticated projects of which some 45% are for the repair, maintenance and refurbishment of the existing stock of buildings and civil engineering facilities. Although less than 20% of all projects can be classified as large, they account for more than 60% by value of all construction work.

As well as major fluctuations in overall demand in the industry as a whole and within each of its sectors, demand from individual clients also

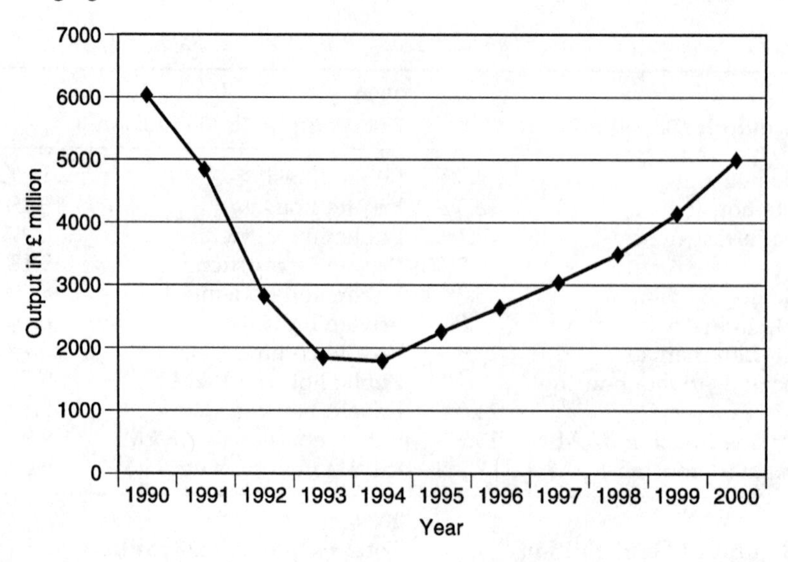

Fig. 1.4. Contractors' output for private sector offices

fluctuates, and in the case of infrequent clients, is discontinuous. It has been estimated that only 20% of construction clients commission major construction work on a regular basis. A full analysis of the complex interrelationship between overall economic performance, government policy and the construction industry is outside the scope of this book. However, it is clear that fluctuating and discontinuous demand does act as a barrier to the development of long-term and more stable and innovative relationships between clients and the supply side of the industry. In turn, supply side relationships also have to be more temporary and transient as supply chains are assembled and aligned to meet the needs of one-off projects and are broken up at the completion of each project. This discontinuous demand means that the supply side of the industry is more geared to flexibility, responsiveness and agility rather than improving ultimate product performance and processes, which may well be more readily achieved in longer-term and more stable processes associated with other industries (Saad and Jones, 1995).

Discontinuous demand impacts on the way construction organisations mange their businesses. When demand is low, organisations find it difficult to fully utilise their workforce. This has been a major factor in the growth of subcontracting. It also contributes to the industry's widely acknowledged reputation for adversarial relationships. There is a tendency for construction contractors to become more cut-throat when tendering for work during periods of low demand and to subsequently recover costs and generate a margin through pursuing claims against the client and/or other project participants. There is also considerable evidence to show that, under these circumstances, many contractors exploit subcontractors by driving down their tender prices, delaying payment, withholding retention and making bogus claims for poor performance. This would suggest that clients have an important role to play in improving the performance of the construction industry by more effective management of their demand for construction products and services and by adopting less confrontational and adversarial forms of procurement.

The number of participants and activities

The large number of participants in the development process reflects the diversity of the industry's clients and products. Clients range from large to small, private to public, regular to infrequent, and from very experienced to inexperienced. Although infrequent clients dominate in terms of numbers, experienced clients account for the majority of construction work. The large number of participants in the process includes contracting firms, specialist contractors,[2] consulting architects and engineers, quantity surveyors, suppliers of building materials, manufacturers of equipment and lawyers. The development process also involves a wide range of activities including establishing the feasibility of and obtaining planning permission for specific developments. Also the regulation, design, manufacture, construction, maintenance, refurbishment and eventual decommissioning of buildings and other structures (Saad and Jones, 1998).

The large variation in the size and complexity of projects has resulted in a large variation in the size and sophistication of construction organisations. Nearly 96% of the directly employed workforce is employed in companies with fewer that 13 employees. More than 90% of all general and specialist contractors employ less than 7 people. This means that

[2] CIRIA (1998) defines a specialist trade contractor as a firm who offers and executes a specialism in any or all of design, manufacture, assembly, installation, testing and commission of items that go into the construction of a building. A specialist contractor offers and executes a design service for the items they manufacture (and/or select and purchase) and install in the building. A trade contractor refers to those who offer work of a skilled nature for the construction of a building but without a design input. Subcontractors are defined as those with responsibility for some part of the construction work (whether with or without a design service) under the employ of a main contractor. This complexity of division illustrates the way in which the industry complicates and fragments the process. The term suppliers is increasingly being used to include main contractors, specialist contractors, trade contractors or subcontractors. This can be seen as a significant step in addressing the fragmentation and division of the industry's supply side.

small firms account for approximately 50% of all employees and produce some 34% of output, much of which is in an informal black economy. Medium-sized private contractors, who account for less than 5% of all firms by number, contribute 32% of total output and 39% of total direct construction employment. The largest contractors, who account for less than 1% of all construction firms, account for 34% of total construction output and generate 20% of employment (CITB, 2002). The larger firms dominate the infrastructure, public sector non-residential building and private sector industrial and commercial construction. Most of their work comprises new construction, where they have captured around 30% of the market, while they undertake only 9% of repair and maintenance works (Drewer and Hazelhurst, 1998).

Organisations have evolved to specialise in different parts of the development and construction process. For each project, a series of decisions has to be made which take into account the context in which the tasks are to be executed and temporary organisational structures have to be created. These individuals and organisations form a temporary and discontinuous chain of internal customers and suppliers. In the case of a complex construction project, this can involve coordinating the activities of several hundred specialist firms. This means that a distinctive characteristic of construction is its fragmentation and lack of integration between the various stages and participants. This is most apparent in the separation of design from construction and the lack of effective feedback mechanisms to inform subsequent design decisions. Under these circumstances, the coordination and integration of information and value flows through the project and supply chains is difficult. These fundamental characteristics of construction explain why it is often associated with a high degree of fragmentation, diversity and interdependency and hence complexity and uncertainty.

Another significant feature in this fragmentation of the industry is the growing adoption of outsourcing by main contractors and the shift from direct employment to subcontracting and specialist contracting. This shift, characterising all industries, is particularly significant in shaping business processes in the construction industry (CIRIA, 1998). The move towards outsourcing greatly increased during the years from the early 1950s to the early 1970s, a period characterised by increasing labour disputes. One method used by main contractors to circumvent such difficulties was the adoption of subcontracting. This approach either transferred labour relations problems to another company or, through the growth of labour-only self-employment, removed the role of trade unions altogether. However, in retrospect, subcontracting transformed rather than removed the labour difficulties faced by main contractors. It increased the number and diversity of organisations involved in projects and main contractors gradually began to lose control over site operations, labour and site labour relations which affected their ability[3] to work together and innovate.

As well as undertaking a diversity of tasks, construction organisations also vary widely in size and, arguably, in their ability to innovative. As can be seen from Table 1.4, the industry is dominated by a large number of small firms.

Even the largest construction firms are small in terms of turnover and asset base in comparison with large firms in other sectors of the economy. The smallest firms are typically jobbing builders and consultants using craft skills and approaches. This diversity and fragmentation cannot be seen as conducive to innovation.

Clients' contracting strategies

Competition is inherent in construction's culture but price is the primary determinant of success. Time plays a role, as does quality, but price most frequently dominates customer-supplier relationships. Contracts are systematically awarded to the lowest bidder and competitiveness is currently largely determined by the supply side's capacity to assemble a temporary network of suppliers capable of responding to the needs of a specific client at the lowest possible price. This emphasis on price competition is seen as amplifying the effects of the adversarial and short-term relationships and the lack of trust associated with transactions in construction (Jones and Saad, 1998).

The Wood report (1975) describes the characteristics of a construction industry characterised by discrete projects and short-term relationships.

> *The seeking of new tenders for each project results in a damaging lack of continuity for the individual firm. At site level, teams of men and managers built up during the progress of a job are often dispersed at the end of the contract. The advantages gained through learning particular constructional operations and sequences, understanding client attitudes and policies, and welding together effective management and site teams are lost ... Furthermore, the incentive for firms to innovate is weakened.*

A further consequence of this approach is that the supply side is pushed into more defensive, more adversarial, claims-orientated attitudes. They are all too often required to enter into lump sum contracts that include a firm completion date on the basis of limited information from architects, clients and engineers. They have to make assumptions about the design

[3] Ability is seen as encompassing the possession of the necessary skills, understanding and power.

Table 1.4. The number of construction firms by size (1990–2000)

Size of firm by number of employees	1990	1991	1992	1993	1994	1995	1996	1997	1998	1999	2000
1	101 223	103 169	94 452	93 585	97 141	99 099	81 363	86 269	87 837	88 018	87 712
2–3	71 498	70 452	68 486	64 438	65 188	64 837	56 106	47 644	47 918	49 350	48 773
4–7	23 403	21 664	30 395	26 072	22 145	20 288	15 317	15 737	16 391	16 969	16 584
8–13	5362	4981	5240	4630	4221	4021	4366	3787	3988	4148	3790
14–24	3935	3429	3574	3129	2881	2828	2952	3101	3274	3271	3104
25–34	1420	1186	1146	1066	956	938	1103	1176	1201	1332	1201
35–59	1305	1100	1148	1098	1008	968	984	1156	1263	1188	1109
60–79	442	382	361	294	325	307	325	396	419	397	364
80–114	432	372	317	283	262	258	263	296	319	304	271
115–299	507	431	387	330	356	337	348	381	405	379	341
300–599	150	137	103	96	92	105	101	107	125	105	91
600–1199	69	58	59	53	50	51	54	60	56	58	51
1200 and over	47	39	36	33	32	33	33	38	40	42	35

that architects and engineers have not had time to finalise. This inevitably leads to disputes when design decisions have to be made during the course of the construction work. Over many years, contractors and other suppliers have learnt how to claim for the effects of extra work, disruption of their normal work patterns and inefficient sequencing of activities owing to late information.

The Reading Construction Forum's report, *Unlocking specialist potential* presents evidence to show that this confrontational approach to procurement has resulted in the following weaknesses in construction that are acting against innovation and thus impeding improvements in its overall performance (Jones and Saad, 1998):

- undue emphasis on price rather than value
- poor communication and ambiguous project information
- adversarial approaches leading to disputes, conflict and a lack of trust between project partners
- insufficient focus on external and internal customer relationships
- lack of clarity of agreed requirements with internal customers and suppliers
- unfair unloading of risk
- fierce price competition and reduced margins.

Much of construction continues to place strong reliance on more and more procurement systems and contracts rather than developing stable relationships based on mutual understanding, appropriate sharing of risk and the development of trust. However, during the 1990s, parts of the industry have begun shifting the emphasis from a contractual to a more collaborative partnership approach (see Fig. 1.5).

This has been led by a number of key participants in the construction process including regular and frequent clients and more recently by main contractors and leading specialist contractors who are well placed to play a unifying and aligning role both upstream and downstream in the processes.

Work environment

A significant feature of construction that sets it apart from many other industries is that much of the building is constructed on a site that belongs to the client. This implies that the main resources of production have to be mobile, moving from workstation to workstation around the site, and from site to site, to undertake a sequence of tasks that, to a degree at least, vary from project to project. The transient nature of construction operations means that projects, irrespective of size, are often

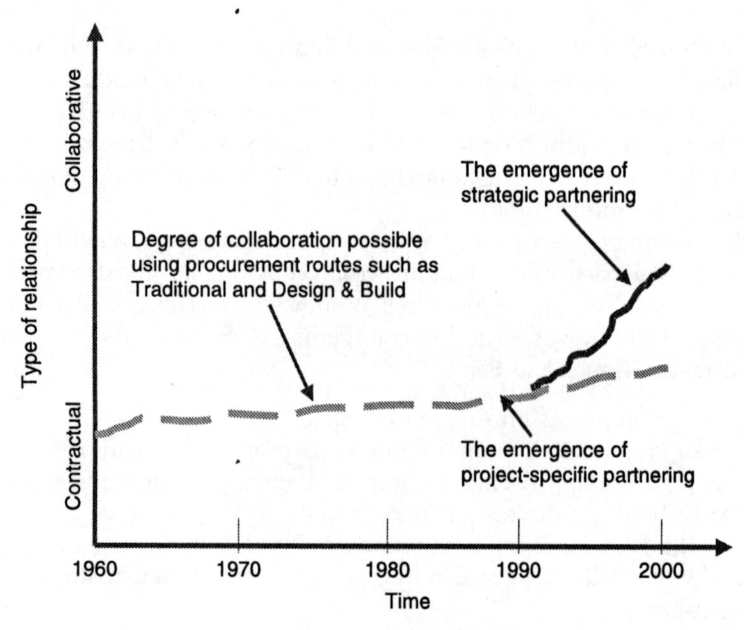

Fig. 1.5. The emergence of more collaborative approaches to procurement based on partnering

hazardous and risky affairs. The dangerous, dirty, unpleasant and often chaotic work environment of the construction site can be seen as a significant factor affecting the industry's poor performance in relation to productivity and quality and contributing to its poor image.

It is sometimes argued, therefore, that the construction process does not correspond directly to the factory environment with its emphasis on Taylorism and Fordism principles. This could explain the limited success of construction in adopting innovation successfully implemented in other sectors during the Fordist era. Once again, this illustrates the difficulties that construction has had in embracing all elements of the factory and mass production organisation. However, with the shifts to post-Fordism, with its strong requirement for adapting the means of production to the market needs, construction's project-based approach could be considered as being well placed to benefit from the new paradigm and improve its competitiveness and performance.

In spite of the failure of the industry to fully embrace Fordist approaches, there have been attempts to replicate factory conditions on the construction site. For example, during the 1990s a number of Japanese companies experimented with creating mobile 'site factories'. These comprised temporary shelters designed to be moved around the building under construction. Work tasks were then executed more easily and effectively within the controlled environment, order and structure of the shelters or 'site factories'. Another example of the industry's attempts to adopt a mass production organisation is illustrated by the large-scale introduction of factory-based production of building components during the 1960s.

Adversarial relationships

The key features of construction can explain why construction is widely acknowledged to have deeply ingrained adversarial attitudes and behaviour. The Latham report (1994) identified a general agreement within the industry that the culture of conflict seems to be embedded, and that the tendency towards litigiousness grew during the recession of the late 1980s and early 1990s. The report suggested that these disputes and conflicts have reduced morale and impeded the development of team spirit. Although acknowledging that parts of UK construction were excellent and the foundations of improvement laid by the Latham report, the Egan report also reaffirmed continuing dissatisfaction with much of construction and its continuing fragmentation. Despite a number of attempts to introduce new procurement methods aimed at reducing the level of conflict, reducing fragmentation and improving performance, the culture remains essentially adversarial, hence impeding innovation.

Risk

The outcome of any construction project in terms of cost, time and total satisfaction of objectives cannot be predicted accurately at the outset. Construction project outcomes are often considered difficult to predict because of the significant number of stages and actors, from both the public and private sectors, involved in the process. Also, as has been seen earlier, construction projects are very often a one-off experience as their characteristics are determined by the requirements and specifications dictated by a specific client. They also involve a wide range of diverse activities. As a result of this diversity, specific projects and the industry as a whole are characterised by considerable uncertainty.

The specific risks involved include the vagaries of the Town and Country Planning system, constant changes in legislation, fluctuating market conditions, limited information regarding ground conditions, the weather, business failures and growing competition. Latham recognised that no project is free of risk and that such risk must be managed, minimised, shared, transferred or accepted. However, in construction, risk is often allocated unfairly and opportunistically, which contributes to perpetuating its adversarial culture. Given the lack of competence trust in relationships outlined previously, there is also a high risk of conflict and the need to resort to litigation.

The fragmentation of projects and the opportunistic behaviour of the participants make it difficult to identify, monitor and manage risk. As a result, construction organisations are reluctant to change and accept the additional risks associated with innovation that are seen as being over and above those already inherent in the development process.

Interfaces between project and organisations

As has already been shown, the design and construction process for even a modest project can involve over a hundred autonomous organisations. A major task in the management of the process relates to the need to reconcile the objectives of projects and organisations. Walker (1984) argues that construction presents two types of management issue – the problem of managing firms and that of managing projects – leading to a complex matrix management structure and limited inter-organisational cooperation. The firms engaged in projects are independent companies but are organisationally interdependent in relation to the project. These organisations often have very different aims, objectives, processes and cultures. These differences also create coordination and communication problems leading to uncertainty and the potential for opportunistic behaviour, lack of trust and conflict between the needs of each autonomous organisation and each project. These difficulties can provide the basis of the benefits to be derived from collaborating and improving relationships, but also explain the reluctance and slowness of many construction organisations to work more closely together and to innovate.

People and training

The construction industry has long experienced difficulties in meeting its skill requirements. The industry relies heavily on middle-aged craftspeople and their average age is rising. Even a small upturn in the industry's activity is likely to result in major skills shortages. Factors contributing to the current situation include the competition for workers from more attractive competitor industries and insufficient numbers of people coming into the industry, particularly in periods of recession such as the early 1990s. In the period from 1989 to 1994 the number of craft operatives fell by 38% from 583 000 to 361 000[4] and the number of new first-year trainees dropped by over 40%, from about 35 000 to 20 000. Also, investment in the training of new and existing workers has often been limited. Although levels of investment in training have been increasing in recent years, there continues to be a major shortfall in terms of properly qualified recruits joining the industry. The Construction Industry Training Board (CITB) report that skill shortages are most acute in London and the South-East and that carpenters and bricklayers are in most demand. These shortages create pressure for mobility of labour between regions and the employment of unqualified workers (CITB, 2002).

Within total construction employment, the largest occupational group are carpenters and joiners (approximately 200 000), followed by managers, electricians, clerical staff and bricklayers. The smallest occupational group was glaziers (approximately 7600). Over the past decade there has been some ageing of the workforce with a decline in the share of 16–24 year-olds and some increase in those aged 45 years and over. The industry has 46% of all those in employment qualified to NVQ Level 3 or above. This is higher than in distribution (31%) but lower than in energy and water (61%). There is currently a persistent decline in applications and acceptances for higher education courses in all construction and built environment-related disciplines, apart from architecture.

Labour-only subcontracting continues to be used mainly to enable the industry to remain flexible. This is seen as both a strength and a weakness. Main contractors can make saving in the costs of employment. It reduces their responsibilities under employment legislation and provides them with greater flexibility in responding to fluctuations in demand. It also can result in self-fulfilment for the self-employed worker and higher rates of pay. However, it also results in avoidance of some tax and insurance obligations and contributes to the shortage of skilled operatives. It also accounts for the inadequate training and investment in new technologies and R&D.

Problems of recruiting women and ethnic minority workers

As can be seen from Tables 1.5 and 1.6, construction experiences difficulties in recruiting its proportion of women and ethnic minorities.

[4] *Housing and Construction Statistics 1984–1994* Great Britain, London, HMSO.

Table 1.5. Employment in the construction industry by gender

	All	Women	Men	All	Women	Men
1990	1265	187	1078	918	23	895
1991	1183	171	1012	837	20	818
1992	1089	166	923	757	18	739
1993	1021	157	864	746	17	728
1994	1025	163	862	797	17	780
1995	996	148	849	802	14	788
1996	1000	160	840	799	19	779
1997	1095	137	958	749	15	734
1998	1217	152	1064	659	14	645
1999	1235	151	1083	664	12	652
2000	1344	166	1178	622	10	612

Figures are from Spring of each year in thousands.

Table 1.6. Employees and self-employed in the construction industry by ethnic origin for the quarter average: Spring 1999/2000 to Autumn 2000

	All	White	Non-white	Black	Indian	Pakistani/ Bangladeshi	Chinese	Other/ mixed
All	1900	1857	43	19	14			
Female	172	167						
Male	1728	1690	38	17	13			

The priorities for construction are to meet the current shortfall of skills against demand, respond to growth in the market, fulfil the objective of a fully qualified workforce and meet the need for higher-level skills and occupational change. This involves maintaining the intake of skilled people, retaining skill levels in the existing workforce and enabling innovation and growth in the industry.

Conclusion

The above review of the key features of construction can help to explain the industry's diversity, wide range of products and services and widely differing sizes of projects. There are large fluctuations in demand for construction over time and with particularly large fluctuations in some sectors, such as offices. Although public sector work is a significant part of construction's output, its significance declined during the privatisations of the 1980s and early 1990s. Given the long life expectancy of construction's products, repair and maintenance work represents around 45% of construction output.

The diversity and complexity of the development process and the diverse skills, materials, organisational structures, technologies and perspectives sets it apart from many other economic activities. The diversity of its sectors also suggests that it should not be seen as a single industry but rather as a group of complementary sub-industries (Drewer and Hazelhurst, 1998). However, it is unwise to overemphasise the uniqueness of construction or specific projects. There are many common features to projects and many of the roles, activities and resources are common to a wide range of different types of construction projects.

This diversity and fluctuating demand within its different sectors means that construction needs to have a high degree of flexibility. This would suggest that it has to be more geared to flexibility and agility than process-based industries.

The industry continues to demonstrate its capacity to innovate as it has responded to changes in markets, institutions, regulatory framework, processes, technologies and in the transformation of relationships between the different participants in the process. However, there is a growing demand, particularly from regular clients, that the industry intensifies its efforts to improve its performance and become a world-class provider of products and services, not only constantly benchmarking itself against its international competitors but also against other sectors of the economy (Latham, 1994).

Construction is, therefore, expected to undergo significant innovation in relationships, processes and technologies in the next 5–10 years as it responds more imaginatively to shifting needs of its clients and addresses the wider issues of social, economic and environmental sustainability. Some of this innovation will focus on extending existing trends such as the transfer of production off site and the use of manufactured and standardised components. Other innovations such as the development of more collaborative relationships, more integrated processes, quality assured products and services and supply chain management represent considerable departures for the industry. As such, they will have consequences for the workforce in terms of learning and the development of different skills and a more collaborative culture.

The next chapters will attempt to examine how the key features of construction have shaped its ability to innovate and change as it responds to these new challenges. They also outline the ways in which construction needs to implement, manage and sustain innovation in ways that match its specific culture, context and specificity yet increases productivity and enables it to compete internationally.

References

BALL, M. *Rebuilding Construction: Economic Change in the British Construction Industry*. Routledge, London, 1988.

BANWELL, H. *The Placing and Management of Contracts for Building and Civil Engineering Work*, Ministry of Public Buildings and Works, HMSO, London, 1964.

BARRIE, D. and PAULSON, B. *Professional Construction Management*. McGraw-Hill, New York, 1978.

BUSINESS ROUNDTABLE. *'Technological Progress in the Construction Industry, A Report of the Construction Industry Cost Effectiveness Project*. Report No. B-3, 1982.

CLARKE, L. The production of the built environment: backward or peculiar? *The Production of the Built Environment*, **6**, 2–7, Bartlett School, University College, London, 1985.

CONSTRUCTION INDUSTRY RESEARCH AND INFORMATION ASSOCIATION (CIRIA). *Specialist Trade Contracting – A Review*. Special Publication 138, 1998.

CONSTRUCTION INDUSTRY TRAINING BOARD (CITB) *Skills Foresight Report*. CITB, 2002.

COX, A. and TOWNSEND, M. *Strategic Procurement in Construction: towards better practice in the management of construction supply chains*. Thomas Telford, London, 1998.

DREWER, S. and HAZELHURST, G. *Construction Towards the Millennium*. University of the West of England, 1998.

EGAN, SIR JOHN. *Rethinking Construction*. Report of the Construction Task Force to the Deputy Prime Minister on the Scope for Improving the Quality and Efficiency of UK Construction. HMSO, London, 1998.

EMMERSON, I. *Survey of the Problems Before the Construction Industry*. HMSO, London, 1962.

GROAK, S. *The Idea of Building*. E & FN Spon, London, 1992.

JONES, M and SAAD, M. *Unlocking Specialist Potential: a more participative role for specialist contractors*. Reading Construction Forum, Thomas Telford, London, 1998.

LATHAM, SIR MICHAEL. *Constructing the team*. Final report of the government/ industry review of procurement and contractual arrangements in the UK construction industry. HMSO, London, 1994.

NATIONAL ECONOMIC DEVELOPMENT OFFICE (NEDO). *Strategy for Construction R&D*, NEDO Books, London, 1985.

NEEDLEMAN, L. *The Economics of Housing*. Staples Press, London, 1965.

SAAD, M. and JONES, M. New Organisational and Cultural Arrangements in the Management of Projects in Construction in *Vision To Reality*. Australian Institute of Project Management, National Conference, 1995.

SIMON. *Report on the Management and Placing of Building Contracts*. Ministry of Works, HMSO, London, 1944.

WALKER, A. *Project Management in Construction*. Granada, 1984.

WOOD. *The Public Client and the Construction Industries*. HMSO, London, 1975.

Chapter 2

A review of the responsiveness of construction to innovation

In order to fully understand today's construction industry and its responsiveness to change it is essential to have some understanding of its past. As mentioned in Chapter 1, a number of reports have identified the strengths and weaknesses in the UK construction industry. Most of these reports, stretching back at least 60 years, have essentially made the same criticisms and identified the same persistent problems. This chapter aims to describe how construction has introduced and managed changes during a major period in its history in order to gain a better understanding of the industry's responsiveness to change, and to identify the main enablers and inhibitors to innovation. It examines the period of industrialisation between the start of the Industrial Revolution in 1760 and the demise of Fordism during the late 1970s.

The factors affecting the nature of construction during this period are shown in Fig. 2.1 and comprise the wider environment, the construction market, construction organisations and their supply systems, the regulatory and institutional framework and the technical support infrastructure. We would argue that the main elements of this model of construction have been significantly influenced and shaped by the development and spread of Fordist ideas and principles.

The rise of Fordism and its impact on construction's wider environment

The wider environment within which the UK construction industry operates was transformed during the rise of Fordist thinking, the origin of which can be traced back to the start of the Industrial Revolution in 1760. During the Industrial Revolution, the world was shaped into the form we now see around us. Great changes took place that altered people's lives and methods of work. Machines powered by water, and later steam,

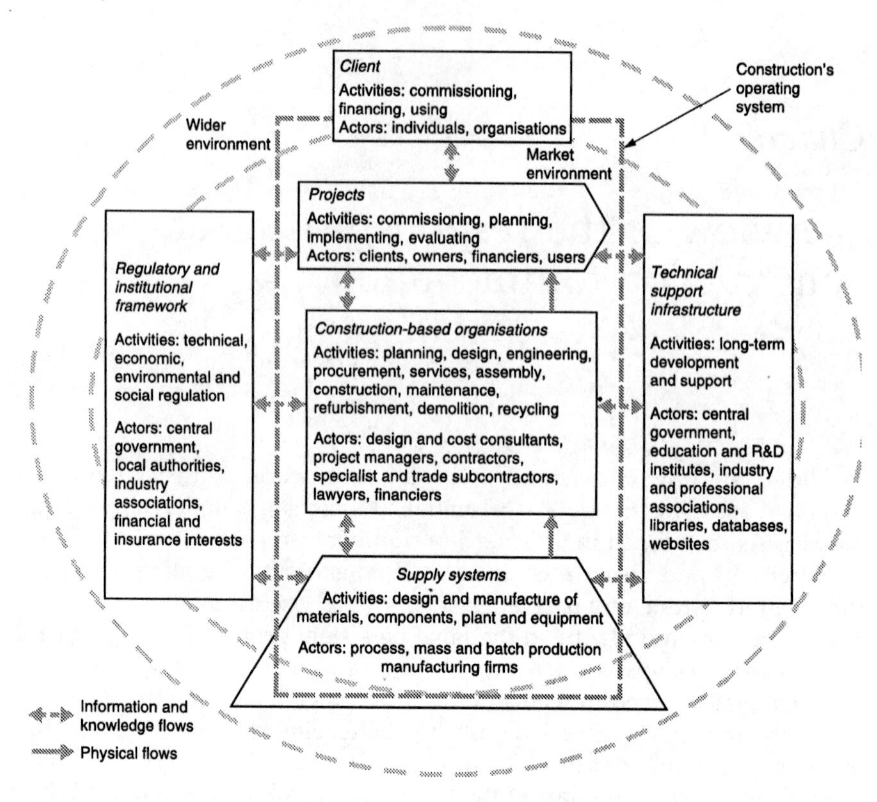

Fig. 2.1. Contextual model for construction (adapted from Gann and Salter, 2000)

were invented to produce cloth and other goods more quickly. Productivity was greatly increased through technology-led innovation. The factory system was introduced as businessmen set up machines in factories, and employed workers to perform a single task in operating them. As it took many workers to run these big machines, there was a massive migration of people from the countryside into the new industrial towns that grew up around mines and factories. Before the Industrial Revolution, most societies were based on small-scale, self-sufficient agricultural production, with the vast majority of the population living in the countryside. After the Industrial Revolution had run its course, the reverse became the case with most people living in urban centres and depending on industrial and commercial activities for their livelihood (Landes, 1969).

The new system of organising work developed during the Industrial Revolution and this came to characterise industrial life in Britain by the end of the nineteenth century and continues to influence the organisation of work today. This was essentially based on the hierarchical and horizontal division of labour, which led to greater specialisation. The articulation and propagation of the division of labour owed much to the work of Adam Smith. In his book *The Wealth of Nations*, published in 1776, Smith argued that a number of people each specialising in one aspect of the production of an artefact was more efficient than it being made by one person doing everything. As the nineteenth century progressed, this approach to work organisation became more developed and systemised. Charles Babbage (1835) argued the need for dividing tasks between and within mental and manual labour and identified three 'classes' employed in the work process:

- the entrepreneur and the technical specialists who would design the machines and plan the form of work organisation
- the operative engineers and managers who would be responsible for executing such plans and designs, based on only partial knowledge of the processes involved
- the mass of employees, needing only a low level of skill, who would undertake the work.

The pioneers of these developments in work organisation tended to be authoritarian and rigid disciplinarians who often used their personal authority to impose the new working arrangements on a normally reluctant workforce (Chapman and Chassagne, 1981). Therefore, change usually needed to be introduced and managed through imposition and force rather than negotiation and agreement. Understandably, there was strong resistance from the workers, both active and passive, to the introduction of new working patterns and methods (Kriedte *et al.*, 1981). This presented an inducement to employers to seek methods of reducing their reliance on skilled labour (Bruland, 1989). One of the ways in which employers responded was through technology-led innovations aimed at producing machinery to replace or reduce their reliance on skilled labour.

This formed the basis of what later became known as Taylorism and Fordism. The key figures in the development of the Fordist approach were Frederick Taylor, Henry Ford, Frank and Lillian Gilbreth in the US, Henri Fayol in France and Max Weber in Germany. This approach was based on a number of basic assumptions. There was only one 'best way' for all organisations to be structured and operated, which was essentially determined by the physical aspects of the technology available. Organisations were perceived as rational entities within which people were motivated solely by financial reward. In addition, human fallibility and emotions needed to be eliminated from the workplace. Workers, described as 'economic men', were seen merely as a continuation of machinery. The most appropriate form of job design was achieved through the use of hierarchical and horizontal division of labour to create narrowly focused jobs. These jobs, which were designed to fit and complement machinery, were controlled by precise standardised procedures and rules leading to the removal of worker discretion and a clear definition of the tasks that workers needed to perform. This allowed workers to be closely monitored and controlled by their direct supervisors (Burnes, 2000).

The main advantages associated with Fordism include economies of scale from an increase in volume of production leading to a reduction of cost per unit of production. It opened up technical possibilities to develop and deploy capital equipment and the opportunity to use less skilled people. It allowed the development of the mass-assembly line, which facilitated the production of high volumes of standardised products made from interchangeable parts with simple fixings and connections. The same system of dimensional discipline was used through the entire process to ensure that components could be fitted together with minimum assembly costs. Unskilled or semi-skilled labour was used to tend expensive dedicated machinery. With the increasing imperative to maintain production, inventory was needed to prevent stoppages of the line. As changing the machinery was expensive, the life of products was extended as long as possible. Although this resulted in customers benefiting from lower costs, they had less variety and customisation of products to their particular needs.

These ideas became codified into a blueprint for mass production that was most successfully used in the early car industry by Henry Ford who, for instance, reduced the task cycle from hours to minutes through the fragmentation of tasks. He also reduced the size and complexity of each task to simple and repetitive operations, easy to master with hardly any skill or qualification. However, this system, characterised by a rigid separation of mental from manual work, attracted a good deal of criticism since too much specialisation led to alienating jobs that were monotonous and hence counter-productive to higher productivity. The worker, who was told exactly what to do and how to do it, was in fact seen as a resource to be operated in a manner as close to that of machines as possible. In addition, by considering the financial reward as the prime motivator, Fordism did not take into account other influences such as the satisfaction to be derived from the act of work itself or from association with others at the workplace, and tended therefore to ignore the fundamental fact that workers were human beings. This means that the assumptions of Fordism, in relation to workers and work organisation, have been perceived by other organisational writers as overly simplistic and generalised (Saad, 1991). The growing awareness of the shortcomings of Fordism led to the development of a new approach, focusing on human relations, which emerged in the 1920s and 1930s as a reaction to the dehumanised scientific approach and its undue emphasis on technology-led innovation.

The human relations approach

The rise of the Human Relations School began with the research conducted by Harvard psychologist Elton Mayo at the Hawthorne plant of the Western Electric Company between 1927 and 1932 (Mayo, 1933). The Hawthorne studies concluded that a worker's behaviour and sentiment were notably related and that money was less influential in determining worker output than group standards, sentiments and security. Production was increased as a result of changes in social relations, motivation, and supervision of workers rather than just financial rewards and technology-led innovation. The worker was no longer considered as being motivated only by financial rewards but also by a series of social and human factors. This brought to the forefront the concept of the organisation as a social system with greater recognition given to human values. Elton Mayo agreed with Frederick Taylor that cooperation and consultation between workers and managers was necessary, but refused to consider employees as 'simple factors of production'. Unlike Taylor, who had suggested that an efficient organisation would create happy employees, Mayo suggested that happy employees would create an efficient organisation.

Recognition of human values has been strongly influenced by the theory of motivation introduced by Abraham Maslow (1954), known as the need hierarchy theory and in which he described the motivating factors that influence the worker's behaviour. Maslow suggested that everyone has various areas of need that exist on an ascending scale, each level motivating the individual, once the need below has been satisfied. The important point in Maslow's theory is that once a need is satisfied, it is not any longer a motivator. This suggests that money is not the only major motivator and that the worker is not, as assumed by Fordism, 'an economic man'. Accordingly, this directed the emphasis away from the satisfaction of basic financial and survival needs towards social, esteem and, ultimately, self-actualisation.[1] Maslow's assumptions, which were essentially about the behaviour and motivation of individuals, could also be related to motivation and behaviour of organisations as a whole.

Another major study of factors affecting satisfaction and dissatisfaction with work was undertaken by Herzberg and his colleagues (1959). In the past, it had always been assumed that the factors causing job satisfaction

---

[1] Basic needs, such as safety and personal development, are far from being addressed in construction as shown by the accident statistics and the need for the Respect for People Initiative introduced in May 1999 when the Construction Minister challenged the industry to radically improve its performance on people issues. This could account in part for the failure of the industry to recruit across the social spectrum, including women. On the other hand, the industry can provide a great deal of self-actualisation through early responsibility.

would, by their absence, lead to job dissatisfaction. For Taylor and his followers, for instance, when the financial reward was relatively substantial, workers would necessarily be satisfied and when the monetary reward was felt to be low then dissatisfaction could result. Similarly, when social and personal needs were fulfilled, job satisfaction would result and when these were not met, then dissatisfaction would rise. Thus, Herzberg's results suggest that different relationships exist between job satisfaction and job dissatisfaction since each situation is caused by an entirely different set of factors. The major factors of satisfaction that Herzberg called 'motivating factors' include:

- achievement
- recognition
- work itself
- responsibility
- advancement.

The set of factors causing dissatisfaction at work and called 'hygiene factors' are:

- working conditions
- company policy and administration
- interpersonal relations
- supervision
- salary.

Dissatisfaction is reduced by factors such as improving wages, by more considerate supervision, and by better physical conditions of work. However, reducing dissatisfaction in these ways will not produce positive satisfaction needed for continuous learning and innovation. This can only be provided by communication, achievement, recognition, responsibility, and by the feeling that a job has been well done. As each state is determined by its own group of factors, the task of management is to address both hygiene and motivating factors.

The Human Relations School also rejected the scientific management structure (specialisation, division of work, many levels of hierarchy and formal control) and its strong dependence on technology-led innovation. It proposed a more democratic, less authoritarian and less hierarchical structure where employees are seen as playing a more important role in production. For instance, McGregor (1960) proposed a democratic and participative approach based on conditions conducive to involvement, commitment, initiative and innovation (Glover and Rushbrooke, 1983; Saad, 1991). Similarly, Argyris (1964) focused on the need for effective relationships between an organisation and its individuals in order to promote motivation and self-actualisation.

The analysis of the major work regarding the development of the human relations approach and organisational behaviour highlights that the employee is not a kind of machine, motivated only by financial reward, but rather a human being whose behaviour is highly determined by his/her psychological perception of work. However, this approach is thought to generalise about employees' behaviour and to ignore personal differences. The human relations approach is also found to be limited to the relationships of employees with their organisation and does not view the organisation as a whole with its internal and external relationships (Saad, 2000). People at work are seen as responding in the same ways to the same set of stimuli. There is a tendency to generalise about people and hence to predict their behaviour in any given set of circumstances. Even in Herzberg's approach, there is, as Glover and Rushbrooke argue, the assumption of a basic homogeneity or 'sameness' of individuals. The way in which 'motivating factors' and 'hygiene factors' are presented suggests that there was a tendency for a set of specific factors to be associated with a specific state. More recent theories of motivation were therefore developed, taking into consideration the way each situation or each individual would react to a specific set of stimuli. The need to take into account individual and organisational goals in motivation and organisation is expanding and becoming more complex and this is why organisation theory has moved towards an investigation of other variables such as people, size of firms and ownership, and their relationships with many organisational features, rather than just technology-led innovation. This has led to the development of the systems and contingency approaches.

The systems approach

In the systems approach, an organisation is viewed as consisting of an external environmental system and an internal system of relationships which are interdependent and which need to be taken into account. This led to the socio-technical system (Emery and Trist, 1965), which views the organisation as a whole, depending on interactions between the internal and external environment and on the integration of human activities and technology. However, the implementation of this global or holistic approach required a change of the basic model in which organisation theory had been conceived. The problem, as Trist (1950) argued, was neither a question of simply adjusting people to technology, nor technology to people, but organising the interface so that the best match could be obtained between both. Thus, any attempts ignoring interaction effects – that is, to optimise one sub-system without due regard to the other – will not lead to an effective understanding of the functioning of organisations. This is reflected in the structure of the model shown in

Figure 2.1, which identifies the key determinants for the development, management and sustainability of innovation in construction and its underlying rationale.

Joan Woodward (1965, 1970) highlighted the relationship between technology, organisational structure and successful performance and suggested the concept of the technological imperative in organisational structure – that is, technology plays a strong role in shaping an organisation's structure and procedures. For this, Woodward used a classification of technology based on operations technology (small batch production, mass production and process production), and found these differences in technology responsible for many differences in organisational structure. It was, for instance, found that process technologies were characterised by taller hierarchies than unit and small batch production. Large batch and mass production were found to include attributes such as a large amount of specialisation, task fragmentation and more paperwork. For Woodward, technology was, therefore, found to be playing a strong role in organisation design but she warned against generalising and specified that 'there is no best way'. The same principle, she added, can produce different results in different circumstances. However, for the Aston group (Pugh *et al.*, 1969), variations in organisational structure could be better explained by other variables such as size of the organisation rather than technology.

Burn and Stalker's study (1966) emphasised the relationships between organisation, technology and environment. The difficulties of adapting to new technology and to market changes, led Burn and Stalker to propose the 'mechanistic' and 'organic' models as two ideal types of management organisation which defined the extreme points of a continuum along which most organisations can be placed. The mechanistic model of organisation is adapted to stable environments and is characterised by repetitive tasks, clear hierarchy of control, vertical communication, and standardised technology. Organisations faced with non-routine activities demanding flexibility and operating within an uncertain environment were identified as organic or organistic models. Burn and Stalker's work introduced some discussion of culture and suggested that innovativeness and flexibility were not simply a consequence of structural arrangements but also related to the underlying culture (Saad, 2000).

The contingency approach

Unlike the traditional organisation theory, the systems approach views the organisation as an 'open system', in continual interaction with its environment (Perrow, 1970) that achieves a 'dynamic equilibrium' (Kast and Rosenzweig, 1985). It therefore provides a new paradigm for the study of the organisation that allows a better understanding of the interrelationships between the key elements of organisation: goals, technology, structure, people and relationships. However, as Kast and Rosenzweig (1985) point out, system concepts give a macro-paradigm, which includes 'a relatively high degree of generalisation'. Thus, more explicit understanding of relationships among organisational variables is essential for improving individual management practice and organisational performance. Indeed, many factors shape organisational behaviour and one approach that has been very successful over the years is the contingency approach that seeks to understand the interrelationships within and among sub-systems as well as between the organisation and its environment. Like the systems approach, this theory also underpins the model, shown in Fig. 2.1, which recognises the interactions between the different sub-systems and their specificities. This avoids generalisations about developing and managing innovation and places a greater emphasis on the interaction of the many variables involved.

With the greater emphasis placed on differentiation, the contingency approach rejects the idea that a particular form of organisation is always best (Lawrence and Lorsch, 1967). In line with this theory, Perrow (1970) developed a model in which he argues that a firm making a standard product using the same processes and procedures will tend to use routine technologies and bureaucratic and mechanistic organisational structures. But firms in non-routine activities dealing with unpredictable environments will use flexible technologies and organic structures. For instance, Galbraith (1973) argues that the more complex and uncertain the environment, the greater the need for a more appropriate organisation with the creation of lateral linkages to facilitate a better flow of communication and decentralisation of decision making to a level where information is located.

Organisations are therefore influenced by a wide range of variables. This indicates that the organisation that can successfully deal with the need for flexibility, uncertain markets and wider environments, non-routine activities and complex technology is more likely to have an organic model. However, there are prerequisites for an effective adoption of the contingency approach, such as a higher level of understanding, experience and skill related to the overall organisational issues rather than just the physical elements of technology.

The impact of Fordism on construction's market

The Industrial Revolution had a significant impact on both the demand and supply sides of the UK construction market. There was growing demand for large factory buildings, where a substantial number of workers were concentrated around the new manufacturing activities. As new industries were set up, people in search of jobs moved from rural areas to housing estates built close to mines and factories, and small market towns grew quickly into great factory-dominated cities such as Manchester and Birmingham. Later, with the use of steam engines in locomotives to pull wagons, tracks including bridges, tunnels, viaducts, cuttings, embankments and other railway infrastructures were needed to support the growing age of railway travel.

It was also a major driver in changing the expectations of the users of buildings. New enterprises and institutions needed new buildings and the mechanical plant necessary to sustain them. People were no longer willing to accept the restrictions imposed by the natural environment on how they lived and worked. Buildings were needed in which an environment appropriate to the internal activities could be sustained throughout the day and night and regardless of the season of the year, and which achieved this as automatically as possible with minimum input from the users.

The initial Industrial Revolution between the 1770s and 1840s, was primarily based on the development of steam energy and mechanisation. Industries rapidly developed to produce the main products of textiles, textile chemicals, textile machinery, and pottery. This was followed, in the period between the 1840s and 1890s, by the emergence of steam power and railways, and industries to produce steam engines, steamships, machine tools, iron and railway equipment. Emerging industries included steel, electricity, gas and synthetic dyestuffs. The period between the 1890s and the 1930s saw the development of electrical and heavy engineering and production capabilities emerged in electrical engineering, electrical machinery, cable and wire, heavy engineering, heavy armaments, steel ships, heavy chemicals, and electrical supply and distribution. From the 1930s and 1940s onwards, the industrial scene became increasingly influenced by Fordism and mass production. During this time, industry was dominated by the production of automobiles, trucks, tractors, tanks, armaments, aircraft, consumer durables, synthetic materials and petro-chemicals. Progress in science and technology, coupled with increasing affluence, gave rise to the optimism that dominated sentiment at the end of the nineteenth and beginning of the twentieth century. However, the first period of the twentieth century was one of spectacular contrasts, with substantial scientific and technological progress on the one hand, but two world wars and the great economic crisis of 1929–33 on the other. Each had significant positive and negative impacts on social and economic conditions in general and construction in particular.

As the economy grew during these periods of industrialisation it needed increasing amounts of energy. Coal gradually gave way to oil as the leading energy source at the end of the nineteenth century, and in the first half of the twentieth century it came to play a major role in construction. Transport and communication improved greatly and the combustion engines made trade and commerce more mobile and global. This provided both new opportunities and new challenges for construction, including the construction of roads and airports and more sophisticated office, industrial and public buildings. The new and increased demand for built assets provided the impetus for increased research, invention and innovation leading to new materials and construction methods as well as improvements in the use of existing materials.

During the periods outlined above, construction became an important sector throughout the world employing many workers as it responded to demands for railways, roads, factories, airports, housing and other built assets. Most building contractors worked in the home market, constructing buildings and other structures in response to changing and growing domestic demand. When the demand for new buildings was drastically reduced in most countries after the 1929 'Great Depression', governments were prompted to launch major labour-intensive projects and some construction firms gained from this through extensive publicly financed infrastructure projects. This was particularly prevalent in the US where contracting organisations such as Bechtel Warren and Kellogg gained strength and, as a result, became well placed to handle complex projects, not only in construction, but also in industrial engineering.

The period between the World Wars saw a further strengthening of construction and its emergence as a major and modern economic sector. The world economy revived to some extent after 1933, and the State became an important client for public works such as roads, railways, military facilities, prisons and offices. Government became more influential as both client and regulator of the industry. As an increasingly influential client, it contributed to the industry's substantial growth during the period from 1945 up to 1973. It was also able to use its growing role as a client as an economic regulator to pursue its essentially Keynesian policy of investment in infrastructure and other projects during downturns in the economy. Through new laws, financial regulation, interest rate policies and assistance for social housing it became increasingly influential as a regulator of the industry.

There are a number of examples where strong demand for construction products and services put major pressure on construction to innovate. Between the First and Second World Wars in the UK, over 4 million houses were built. The end of the First World War made the provision of housing politically important as the optimistic post-war government promised 'Homes for Heroes'. Such was the demand for housing, the Housing and Town Planning Act was passed in 1919 to introduce subsidies for working-class houses built by local authorities. The housing shortage during this period prompted the use of some new, more industrialised construction methods for house construction. Although this demand extended over two decades it produced little innovation in the development of non-traditional methods despite the waiving of by-laws preventing the use of novel methods and materials. Indeed, as reported by Bowley (1966), an Inter-departmental Committee on House Construction covered only 54 400 non-traditional houses built before 1943 of which about 50 000 had been built by the end of 1928. Also, these limited innovations were not sustainable and were largely abandoned during the late 1920s with only the prefabricated systems surviving until the Second World War. Most of the non-traditional houses were built for local authorities who, by 1930, had practically given up building them.

New ideas in relation to the nature of the built environment also influenced construction products and processes. For example, following the period of the 'liberty' style or 'art nouveau' form of architecture, more functionalist forms of building began to emerge. This led to more geometric forms with few of the decorative elements of historic styles. Several other styles emerged which became amalgamated into an 'international' style. All of these changes were made possible by use of new materials, forms of construction and production methods increasingly available within the industry. For example, the use of curved forms of construction using concrete gave birth to romantic structures, and new forms of cladding allowed greater variety in the external appearance of buildings. Internal changes were also demanded and were satisfied by new, more flexible ways of dividing up space and providing more sophisticated services within internal coverings and finishings.

The 20 or so years after 1945 provided another period of improving economic conditions, technical progress and optimism. The ingenuity of scientists and engineers shifted from the war effort to peacetime innovations including telecommunications, computers and materials technologies. There was an immense increase in productive capacity in goods and services and knowledge. Vigorous post-War enterprises and institutions needed much larger, taller and exciting new forms of buildings, and the more sophisticated plant and equipment necessary to support and sustain them. People became more affluent and the standard of living for most people improved. There was high demand for new and better housing, schools and other buildings to satisfy the changing needs and rising aspirations and to rebuild the country's infrastructure following the devastation of the Second World War.

This increased and changing demand in the market stimulated new construction technologies and the emergence of a recognised construction industry with new forms of project-based firms, organisational structures, operating system, regulatory and institutional frameworks and technical support structure as shown in Fig. 2.1. As well as domestic demand, international work allowed some of the larger UK contractors to become multinationals. Even so, construction firms remained smaller and more focused on domestic markets, on average, than the leading firms in other sectors such as manufacturing. The rapid growth in both construction demand and output from the 1950s to the early 1970s provided conditions favourable for the development and use of new technologies and methods. This included further attempts to change from traditional labour-intensive methods to new production methods. This shift in approach became known as the industrialisation of construction and is examined in more detail in the next section of this chapter.

The changing nature of the construction market and its extensive projects in housing, industry, commerce, transport and city development from 1945 onwards formed the main driver of the development of a modern construction industry up until the early 1970s, when it was dealt a severe blow by the first energy crisis in 1973. This resulted in the rapid increase in the cost of energy and the need for greater energy conservation in buildings. As interest rates and the cost of borrowing soared, clients demanded shorter design and construction periods for their projects. During this period there were also shifts in the sociology of work itself leading to a reconsideration of the Fordist approach to the organisation of production with its emphasis on specialisation and division of work. This, in turn, impacted on the design of buildings for industrial and commercial purposes which reflected the concept of 'small is beautiful' and a desire to return to smaller types of organisations. There was also growing dissatisfaction with the quality of much of the built environment including the public housing provided by many local authorities.

From the early 1970s onwards, the growth in demand for construction declined and deteriorating profit performance affected the structure and organisation of the industry. Real average rates of return on capital employed, adjusting for inflation, were almost halved between 1971 and 1981, the burden falling hardest on medium-sized firms, squeezed by the proliferation of very small firms and by the large contractors undertaking smaller projects. The number of firms with seven or fewer employees doubled and sole traders increased by 150%, while firms with more than 600 employees halved between 1975 and 1985. Construction, like other sectors of the economy, was pushed by significant external factors to review the nature of its products and processes. This could be linked to the decline of the fourth Kondratieff long wave and emergence of the fifth wave, which is explained in Chapter 4.

From this outline of the construction market from 1760 to the late 1970s, the Industrial Revolution can be seen as a major factor in shaping the industry and stimulating innovation. The ways in which the industry responded to market changes are outlined in the following section.

Construction's response to market changes and opportunities in the wider environment

As a result of the industrialisation outlined in the previous section, buildings and other structures were needed more quickly and economically to meet the changes in the construction market and the wider environment. The pre-Industrial Revolution ways of building could not deal with increased and changing demand in the market and new technologies, and new forms of organisation became necessary. The industry had to develop a new and more efficient operating system to deliver more complex and sophisticated products capable of matching the increasing expectations of its clients. The regulatory and legal framework had to change in order to curb the excesses of development but at the same time allow for the introduction of new technologies and methods. A more sophisticated technical support infrastructure emerged to support the key actors in construction. During this period these increasing requirements came to be satisfied in buildings of all types and ways of working. However, there is considerable evidence that change was at a slower pace in the UK than in a number of other countries.

The technology of construction changed with new materials such as steel, concrete and plastics coming into use, and existing materials such as glass and iron being improved and produced in new industrialised ways. There were also significant innovations leading to the introduction and use of new construction materials and technologies including the introduction of iron and then steel, reinforced concrete and a steady increase in the use and sophistication of building services. Prefabrication came into greater use with more components of buildings being produced in workshops and yards and later in factories for on-site assembly. Yet, despite the industry's considerable achievements during this period there were continuing and persistent complaints from clients and the government of inefficiency, poor standards, excessive costs and even corruption.

As well as stimulating demand for new construction products and services, the introduction of new sources of power and industrialisation led to innovations in the way building materials were prepared. Most building materials have to be subjected to various preliminary treatments to convert them to a more usable form including casting and moulding, carving, cutting, drilling, grinding and bending. New kinds of machines, such as the milling machine for producing flat surfaces, were developed. The advent of steam power and later electrical power made possible new and less labour-intensive methods of forming building materials into more usable standard shapes and on a much greater scale.

Bowley (1966), Guedes (1979), Sebestyen (1998) and Morton (2002) have all undertaken studies of the evolution of these materials and new technologies, which provide insights into the ability of construction to adopt innovations and hence gauge the receptiveness of the industry to change. It is outside the scope of this book to undertake a detailed analysis of all of the technological innovations during this period. However, a number of examples of innovations are described in order to assess construction's receptiveness to change. Bowley (1966) and Guedes (1979) provide analyses of an early example of innovation in construction – the development of cast iron, wrought iron and later mild steel. This provides a useful insight into the innovativeness of the UK construction industry during the early years of the Fordist era. There were three types of iron and steel that became successively important in the building industry, which made new structures such as bridges, railways and building frames possible. Although the use of cast iron as reinforcement and ties in brick and stone structures dates back to antiquity, engineers became particularly interested in its use in building during the eighteenth century. In 1706, Christopher Wren used slender cast-iron columns to support a gallery in the House of Commons. The first successful iron bridge was erected over the River Severn at Coalbrookdale in Shropshire in 1779, and by 1800 iron bridge building was firmly established in Britain. As the demand for wider spans grew, engineers invented many new techniques to exploit the tensile strength of the material. Many other structures were built during the nineteenth century in which cast and wrought iron were used in many ways. In the case of industrial buildings, there were numerous innovations in fireproofing and beam, column and floor design. Innovations in roof construction were made possible through the development of a wide range of different truss-roof shapes that combined cast iron with timber, timber with wrought iron, and finally wrought iron on its own. Railway stations, especially in most large cities, provide visible examples of the way large spaces were covered with iron roofs using innovatory designs.

Commercial buildings using iron became widespread in the nineteenth century and by the end of the 1830s many shop fronts and beams above large shop frontages were being constructed using cast iron. Such cast-iron fronts were also exported to the expanding colonies and other markets. Public buildings and places of assembly and entertainment were among the first building types to employ iron structural elements, and Foulston's Theatre Royal in Plymouth (1811–1813) was probably the first public building to employ cast- and wrought-iron framing on a large scale. Other types of smaller iron buildings, which were made for export around the world, became extremely popular after the 1830s.

The Crystal Palace, designed by Joseph Paxton, was a highly influential and innovatory structure in iron. It was built for the Great Exhibition of Industry of all Nations held in Hyde Park, London in 1851. The Exhibition nearly failed to get off the ground as the great architects and engineers of the day charged with designing a large temporary structure, came up with a brick monstrosity, which was judged impossible to build in the limited time available. Paxton's design was remarkable not only for its scale, but the speed and method of its erection because it was completely prefabricated using iron, glass, and wood components. There were, however, drawbacks with the *Builder* snidely reporting that 'On the last day of the Hyde Park Exhibition rain came through the roof like a cullender'.

Despite its considerable success as a new building material, cast iron has two inherent weaknesses – its weight and brittleness. These problems were addressed in 1854 when Henry Bessemer accidentally developed a process of converting pig iron into steel. By varying the proportion of carbon in iron, he found that the properties of the resultant steel could be carefully controlled so that it could be made highly elastic or ductile or hard and durable. Later, it became possible to give steel many other properties by alloying it with other metals. However, the substitution of steel for iron in building structures in the UK was much delayed by engineers who were excessively critical of its merits. Its adoption was also impeded by their lack of understanding of the advantages of standardisation of sections and practice (Bowley, 1966).

Indeed, it was nearly 20 years after the Home Insurance Building had been acclaimed in Chicago that the first steel-framed building, the Ritz Hotel, was started in London. Even then, the designers did their best to make it look respectable by concealing all signs of the frame. Bowley (1966) argues that it was the Royal Institute of British Architects (RIBA), the professional body representing the architectural profession, which was primarily responsible for the slow adoption of steel-framed buildings in Britain. The RIBA was very cautious and was of the view that builders needed to be controlled by definite rules, which should only be made after careful discussion. She argues that the architectural profession had not made up its mind sufficiently about steel-frame construction, even though it had been in use for practically a quarter of a century elsewhere, to enable it to advise local authorities on its use before 1909.

This slowness to adopt the new material should not be attributed solely to architects, as the UK building industry as a whole was not particularly interested in steel-frame construction. The economies of steel, although substantial compared to the cost of composite construction, were not as spectacular for the low to medium-rise buildings constructed in Britain as for the much taller buildings constructed in the US. There was also no particular desire on the part of clients in the UK to replace traditional heavy solid walls with lighter walling materials, such as glass. There was also a lack of interest in using steel from other key actors in the industry. The builders considered that frames were more suitable for erection by civil and structural engineering firms. The position of the engineers is more uncertain, but Bowley argues that there are few references to the use of steel in the *Proceedings of the Institution of Civil Engineers* during this time. She goes on to argue that if the engineers were sceptical it is not surprising that architects, with less technical knowledge, were unwilling to use the new material.

Bowley (1966) further argues that another factor in the industry's reluctance to embrace the new material was the UK steel industry's lack of constructive interest in the building market. She contrasted this with Germany where the Ironmakers' Association combined with the association of Architects and Civil Engineers to encourage the use of steel in building, reduce its costs, and assemble all available information to advance the new techniques in building. In addition, the much more rapid acceptance of steel in the US industry was partly attributable to the long cooperation between many of its steelmakers and the American Society of Civil Engineers in resolving the early problems associated with the adoption of structural steel. Moreover, both in Germany and in America steel producers had already expanded the scope of their businesses into structural engineering themselves and made considerable contributions to the design of structures. This greater understanding helped standardisation, which contributed to reductions in cost and explained its economic advantage. In contrast, similar integration between British steelmakers and structural engineers did not begin at all until the end of the 1890s. This shows a more considered, systematic and integrated approach to the development and marketing of new products and a more receptive attitude towards innovation by many actors in the design and construction process in Germany and America.[2] These German and American cases provide early examples of the need for a holistic and systematic approach to innovation in construction that embraces social as well as technical factors and all parts of the industry. Successfully adopting, implementing and sustaining innovation in construction requires a coordinated response from the key participants – clients, construction-based organisations, and supply systems. It also needs the involvement of the regulatory authorities and support infrastructure as illustrated in Fig. 2.1.

This early period of the Fordist era also saw attempts to change the products of the construction industry to adopt prefabrication. The use of mass-produced cast iron components was advocated by William Vose Pickett in a number of pamphlets published in the 1840s and 1850s. It was used successfully, if untypically, in completing the Crystal Palace for the 1851 London Exhibition in less than six months. Other architects and designers, such as Owen Jones (1809–1874), were strong advocates of

[2] A more in-depth analysis of the different types of innovation and their main determinants is provided in Chapter 4.

prefabricated moulded elements produced in large numbers. Jones played a part in constructing a number of buildings in cast iron including an ornamented kiosk that was exported to India in 1870. Prefabrication also became more significant in relation to design with the standardisation of major structural columns and beams of cast iron. However, there was also considerable resistance to the concept of standardisation and prefabrication. For example, John Ruskin (1819–1900), another influential architect, strongly objected to the use of mass-produced moulded elements on the grounds that they degraded the nobility of craftsmanship. Despite this resistance, from the mid-1840s onwards, iron founders and other contractors in Britain expanded their markets to cover the supply of railroads, manufacturing plant, public utilities, complete buildings and prefabricated components throughout the world, but especially in the rapidly developing areas of South America, Asia, Australia and Africa.

The housing shortage following the First World War provides a later example of the UK construction industry's responsiveness to change. This demand for housing, coupled with the shortage of labour and traditional construction materials following the war, led to early experiments in industrialised methods for house production. As discussed earlier, the demand for housing after World War I led to the passing of the 1919 Housing and Town Planning Act which introduced subsidies to local authorities for non-traditional, working-class houses. Such was the pressure for more housing that an Interdepartmental Committee on House Construction was formed to promote innovation in construction. This is an early example of the regulatory and institutional framework exerting considerable pressure on construction to change. However, despite the high demand for housing over two decades, a considerable number of suggestions for new methods, substantial support through the regulatory and institutional framework, and developments in the technical support infrastructure, there was still little sustainable innovation in the development of non-traditional methods. Indeed, only some 54 400 non-traditional houses out of a total of 4 000 000 houses were built before 1943, and of these about 50 000 had been built by the end of 1928. Most of these non-traditional houses were built for local authorities who, by 1930, had practically abandoned them. This first experiment in introducing more industrialised methods shows that, although certain elements of the construction model shown in Fig. 2.1 were in place, the innovations were not sustainable, and were largely abandoned during the late 1920s with only the prefabricated systems surviving until the Second World War.

A number of reasons have been put forward for the failure of these new methods to significantly penetrate the UK construction market. Bowley argues that the influence of architects in discouraging the use of non-traditional methods by local authorities cannot be ignored.

Although local authorities also preferred to delay building housing until traditional material production was restored to pre-war levels rather than accept the additional risks associated with non-traditional methods, she also argues that prejudice and ignorance played a part. The vagaries of housing policy was also a factor as the original housing policy was abandoned in 1921 and not effectively replaced until 1924. This delay allowed additional time for brick production to recover and increase its production. In 1927, the government considered that supplies and costs had eased to the point where housing subsidies could be reduced, further easing the pressure on innovation. Also, innovations in the brick industry between 1930 and 1935, including the application of Fordism and mass production methods, made it much more difficult for the fledgling non-traditional methods. This illustrates that although applying Fordist principles was difficult in construction's operating system it was increasingly successful in the industry's supply systems in the mass production of basic materials such as bricks. Finally, and perhaps most importantly, the new methods did not provide clear competitive advantages over traditional methods and in some cases were clearly inferior.

Although full-blown industrialisation proved impossible, there were some less radical innovations that were highly successful and came to be widely used in house construction. These included the development of the so called 'breeze block' which was a direct response to brick and bricklayer shortages. These were used for internal partitions and as a cheaper alternative to brick in the inner skins of cavity walls. Plasterboard and concrete tiles were also developed but these did not come into full use until after the period of experimentation and their adoption by private house developers in the housebuilding boom of the thirties. These examples show how early attempts to introduce radically new methods of construction in the UK were unsuccessful. However, they did lead to substantial innovation in the supply systems and some smaller yet significant innovations in the operating system leading to some rationalisation of traditional methods.

The first half of the twentieth century saw the emergence of thinking which was to lead to a later test of the innovativeness of the UK construction industry: the Modern Movement in architecture, whose proponents including Le Corbusier, argued the backwardness of construction thinking (Ball, 1988). In arguing against the complexity and aesthetic individuality of traditional building design they set the scene for the second major experiment in industrialisation of housing and other forms of building during the 1960s and 1970s. They maintained that buildings should be designed on the same criteria as the other products of industrial societies and produced in essentially the same way. Writing in 1930, Le Corbusier said:

The task is to define and apply new and clear methods enabling the making of useful dwelling plans which lend themselves for execution in a natural way for standardisation and industrialisation.

The proponents of this movement argued that the main reason why the construction industry had failed to advance at the same rate as other industries could be attributed to the complexity and ornateness of traditional building designs, which tended to place undue emphasis on the aesthetic individuality of buildings rather than their functionality (Banham, 1976). During World War II the 'Architectural Forum' ran a series of articles called 'Building's Post-war Pattern', which called for radical change in construction. It pointed out that:

In no other field of economic and social endeavour is production so involved, assembly so disorganised, distribution so devious. In no other field is there so large a number of separately functioning organisations between raw material and ultimate customer, such haphazard diversity of product, price and performance.

Clearly, construction was seen as being at odds with the development of science, Fordism, and the mass production model which so dominated manufacturing. This model placed a greater emphasis on:

- the production of standardised commodities
- the expression of beauty through function
- continuity of production implying a steady flow of demand
- integration of the different stages of production
- a high degree of organisation of work
- mechanisation to replace manual labour wherever possible and
- research and organised experimentation integrated with production (Stone, 1976).

Buildings, it was argued, should be designed to the same criteria and in the same ways as other products produced by the mass production model.

The years following the Second World War provided an opportunity to embark on the UK's second major experiment with non-traditional methods of house construction. This second period of experimentation was influenced by Fordist principles as the industry attempted to achieve similar benefits to those available in manufacturing and to shift from traditional, accepted practices to new methods based on analysis, research, development, manufacturing and marketing. As in the period following World War I, there were shortages of traditional materials, of bricks and timber and all types of site labour as well as bricklayers. New methods were sought which would provide alternatives less reliant on the traditional materials and site labour, particularly bricklayers.

This second experiment was again strongly influenced by government policy as it closely controlled building during the post-war period up until 1954. Policy in general was aimed at rapid expansion of capacity of the building materials industry. Following a period where the expansion of the industry was stopped, in a change of housing policy at the end of 1950, housebuilding was encouraged again but without an expansion of the building industry. Local authorities (LAs) were once again encouraged to make fuller use of non-traditional housing which had fallen to about 19 000 homes a year. In 1951, the policy changed again and firms producing non-traditional houses were urged to double their output and the government undertook to guarantee such houses for loans and subsidies, provided they approved the plans. The housing drive of the early 1950s saw central government issue volumes of advice and directions to local government on efficient housebuilding. This led to a surge of experimentation in industrialised or system building. As a result, the output of non-traditional housing had increased to 39 400 by 1954 although it still only accounted for just under 20% of output of LA housing. There were also major regional variations in the adoption of the new methods with 45% uptake in the South-West and only 10% in London.

Prefabrication was a key element of this second experiment in industrialisation. Prefabrication was not new to the UK, as the British were exporting prefabricated buildings in the 1870s and the rapid construction of the Crystal Palace in 1851 had already demonstrated its advantages. As in the first experiment, there were no shortages of suggestions for non-traditional methods of house construction. From 1945 to 1947, the emphasis was on using redirected resources no longer needed for armament production for the design and production of temporary prefabricated bungalows. Tens of thousands of houses were built using non-traditional systems including the Portal House, the Aluminium Bungalow, Arcon and Uni-Seco, although these systems were abandoned in the 1950s. By 1948, the 'Third Report on House Construction' provided a list of 561 sponsors of methods that were worth trying out as prototypes. An analysis of the make-up of the sponsors is enlightening and provides a view of the main clusters of innovators and the continuing fragmentation of the industry. Twenty of the sponsors were builders and contractors, five were engineering firms, fourteen firms had a major supply system interest in terms of a particular material or item of equipment, two were architects, four were LAs, one sponsor was classed as miscellaneous and the background of fifteen was unknown. This list of sponsors illustrates the fragmentation of the industry's operating system at this time. In terms of materials and forms of construction, the majority of builders proposed various types of concrete systems of construction and concrete houses.

In 1964, there were over 30 prefabricated systems for low-rise housing, 15 for high-rise and 15 for education and health buildings. Some were developed in the UK but a number were imported from mainland Europe. This new approach to building was enthusiastically embraced by central government, local authorities, planners, architects, and contractors. However, many of the new methods disappeared without even being tried. Nine of the builders did not persevere with their schemes beyond the prototype stage because of a variety of reasons including their unsuitability on technical grounds, their unattractive appearance and design of internal space, uncertainty of maintenance costs, excessive requirements for skilled labour and the difficulty of obtaining steel.

At the peak of prefabrication and industrialisation in 1967, over 70 000 dwellings were built using such large panel systems, but by 1980 this had slumped to virtually zero. With the decline in local authority housebuilding programmes following the privatisation of the 1980s, coupled with the unpopularity of the use of high-rise construction, the use of other systems also declined. The housing associations, which took over the development of social housing from local authorities, and private housebuilders continued to use traditional methods of house construction (Morton, 2002).

Various industrialised systems for school building were also developed in the UK including the CLASP (Consortium of Local Authorities Special Programme) in 1957 and SCOLA (Second Consortium of Local Authorities) systems. Both systems were based on light steel frames, economical to a height of four storeys, to which various prefabricated cladding elements could be fixed. They were designed on a modular basis so that components could be fitted together in a variety of different ways. CLASP was used for office buildings and housing as well as schools up to two storeys high and, if cleverly designed, could be used to produce architecturally satisfying buildings. CLASP and SCOLA were successful in satisfying peak demand but as soon as this slumped the systems were again largely abandoned (Morton, 2002).

Attempts to introduce industrialisation and prefabrication were made for a number of reasons. Shifting much of the production to the factory was seen as a way to address housing and other building shortages more economically. It was seen as a way of embracing the dominant Fordist paradigm and provide better and more productive working conditions for construction workers. It was also seen as a way to address construction's continuing poor reputation for quality and reliability and controlling and coordinating the whole design and construction system. Indeed, for a time, industrialised 'system building' became popular, particularly in the form of large-panel concrete housing. In these industrialised building systems, components and sub-systems were integrated into an overall process with the key elements of industrialised factory production, road transportation and mechanised site assembly techniques using less skilled labour.

Such was the enthusiasm for these systems, it was some 10 to 20 years before it became obvious that although the large-panel systems were capable of producing buildings in large numbers, the inherent restrictions imposed by such systems required too many compromises and the enthusiasm for high-rise industrialised housebuilding declined as quickly as it arose. The emphasis on maximising the use of plant and equipment needed for their production compromised the needs of the end users and the satisfaction of aesthetic and functional considerations. As a consequence, flexibility was reduced in planning individual buildings, and future changes to systems buildings were also difficult. The adoption of prefabrication on such a scale required close liaison between the client, architect-designer, the manufacturer producing the prefabricated components, the manufacturer and supplier of the mechanical plant required for transportation and erection, and the main and specialist subcontractors for their assembly on site (Davey, 1964) – all of which were difficult in construction's highly fragmented operating system and industry structure. Also, in the case of high-rise developments, there were serious social problems as communities were broken up and decanted into faceless, repetitive, mass-produced units. Ball (1988) argues that the state saw industrialised systems made fashionable by the Modern Movement as a way of cheapening the provision of working-class housing and a means of weakening the building unions. He goes on to argue that there was little concern over the effects on end users who ended up paying the cost of the experiment in poor quality housing, characterless office blocks and a barren urban landscape. The final nail in the coffin of the industrialised housing programme was the partial collapse of a block of system-built flats known as Ronan Point in 1968. Three people were killed and twelve injured, and the following enquiry revealed a catalogue of mistakes in the design and construction of the load-bearing panel structure.

Although Gann (2000) argues that there has been no systematic measurement of the overall gains resulting from the use of prefabricated components, evidence suggests that industrialised building systems applied to housing did not raise overall productivity or reduce costs in comparison with craft approaches, or indeed substantially reduce the time taken to construct buildings. When the programmes for public new housing declined in the 1970s and 1980s, the application of these systems was largely abandoned and because of the problems identified above, the idea of industrialised building on this scale fell into disrepute in the UK. This suggests that the economic and social cases for industrialisation had not been clearly made. Throughput of production was never sufficient to justify the massive investment needed to fully embrace Fordist principles. The integration of all elements of the model shown in Fig. 2.1 proved difficult. The systems based on heavy precast concrete panels called for heavy investment in plant at the casting factory, in transport to move the large, heavy panels to the site and lift them into their final position within the building. This meant that the economies of scale from industrialisation and the mass production model attained in other sectors of the economy were never achieved in construction, as it was always a relatively minor part of the overall construction market. On the other hand, the lightweight prefabricated systems used in the school building programme were largely successful in solving acute accommodation problems in education in the 1960s during a period of tight budgetary constraints.

There were also considerable technical problems that were not quickly or adequately addressed. Making joints that were adequate both structurally and in excluding wind and precipitation proved problematic. These led to defects such as water penetration, and inadequate levels of insulation and ventilation led to condensation. Gann (2000) argues that the workers and organisations dealing with such defects were often not part of the original production team. This meant that their knowledge of the technical problems caused by bad design and workmanship was often not captured and fed back to the design and construction team. Another factor was that many of these firms were small and medium enterprises (SMEs) without the resources to invest in training and in research and development. Although prefabrication reduced the number of components, there were also problems associated with ensuring the accuracy and fit of such large construction components by unskilled operatives under site conditions. While labour time was saved in the erection of prefabricated components on site, such buildings often needed to be completed by traditional finishing trades, which reduced much of the earlier productivity and gains. This meant that construction continued to be influenced by craft approaches and high production costs which did not drop with volume.

Clearly, much of construction experienced difficulties in fully embracing industrialisation and systems building, which resulted in two distinct development paths for the industry – craft and industrialised methods. This meant that both paths of development continued along their respective trajectories up to the end of the Fordist era and even to the present day, with both systems having strengths and weaknesses. The industrialised systems largely failed to deliver competitive advantage or the range of choice users wanted. On the other hand, craft processes remained too time-consuming, costly and incapable of delivering the kinds of more highly engineered buildings and structures required. It could be argued that through rationalised building the industry has avoided making the choice between craft or industrialised techniques and attempted a synergistic but extremely challenging combination of the two traditions. This experiment in the radical industrialisation of the industry also impacted on the operating system. Craft skills were eroded as tasks were divided and subdivided, and the control traditionally exercised by craftsmen was replaced by greater specialisation and division. As well as the workforce becoming deskilled, employers became more casualised with the emergence of labour-only subcontracting. Large contracting firms and the suppliers of materials began to monopolise parts of the industry following the path of industrialisation and work tasks became more narrowly specialised.

Although the above examples show that construction failed in fully, or indeed successfully, embracing industrialisation and the mass production model, Morton (2002) argues that it is easy to underestimate their impact on construction. While radical industrialisation was not possible, Fordist principles went on to have a major impact on the industry. Despite the failure of fully integrated building systems, the advantages of prefabrication remained clear and smaller precast concrete components continued to be produced. Prefabrication continued to be used for certain appropriate parts of buildings including doors and windows, kitchen and bathroom units and appliances, partitions and facades. Preformed structural frames in steel, and to some extent concrete, remain common and the continuing prefabrication of cladding, interior wall panels, toilet pod modules, modular lifts, plant rooms and other elements of buildings all bear testimony to the influence of Fordist principles and practices on construction. Alongside these changes, however, most buildings continued to be designed and constructed using traditional or rationalised traditional building methods. The latter refers to a method of building in which organisational techniques associated with Fordism are applied to the erection process without involving a radical change in the forms of construction or necessarily in the techniques of production (Stroud Foster, 1973). Rationalised building seeks to achieve a properly integrated system of design and production leading to continuity in all production operations. Standardisation, prefabrication of components and mechanical plant are all used in the most appropriate way contingent upon the particular circumstances. In this approach it is still essential that the design of the building and the production operations are considered together during the early stages of the project – a way of working, as can be seen in the next section, that much of construction has found it difficult if not impossible to develop.

Many of the changes to construction's products depended on a number of concepts, such as dimensional coordination – a key prerequisite to the industrialisation of construction. Although some form of dimensional coordination had been practised for a long time, the increased use of larger prefabricated components led to the need for more sophisticated and reliable forms of modular and dimensional coordination. Modular coordination has continued to assist the building industry and its supply systems in ensuring that components and sub-systems fit with each other and that their assembly can take place without cutting and that components are interchangeable.

Another major prerequisite to the introduction of new approaches to building after 1945 was the concept of building performance. This has become accepted practice in formulating and checking the functionality of buildings. Traditionally, buildings were required to provide shelter, space and be aesthetically pleasing by using a limited range of materials and forms of construction. Traditional building developed from the use of forms of construction evolved by the traditional building crafts, particularly those of bricklaying, carpentry, plastering and roof tiling and slating. With the proliferation of materials and techniques as a result of industrialisation, the idea of buildings meeting carefully determined user requirements emerged. The definition of the performance of the building can only be realised after the user's requirements have been identified and quantified. This led to the performance concept, which enables designers and contractors to achieve a stated standard of performance by using the most appropriate materials and designs, evaluate alternatives and consider innovative solutions. At the same time, if the performance system is guaranteed by its providers, the client, initial and future occupants, and wider society are protected from undesirable surprises during the life of the building (Sebestyen, 1998) .

As well as changes in construction products during this period, there were significant changes in the technologies and methods of production. Building is essentially a process involving the preparation of a site, bringing in materials and components, forming and assembling these into elements of the building, installing services and finishing ready for hand-over. Over the centuries, the transportation and lifting of heavy building materials had required the physical effort of many labourers and animals and the use of often quite complex equipment, tools, scaffolds, ropes and winches. By the middle of the nineteenth century, general builders had extensive workshops with steam-driven machinery for the working of stone and timber. At the same time, steam engines were being applied from early stages in their development to lift heavy objects on sites. Other devices driven by rotative steam engines from this period include cranes, excavators and pile drivers. The steam mechanical digger or shovel originated in the US in the 1830s and a steam dragline-excavator was patented in France in 1859. As the steam engine could find only limited use on construction sites, the large-scale mechanisation of construction took a major step forward with the invention of the internal combustion engine towards the end of the nineteenth century. This invention and the electric motor radically changed construction methods.

Mobile concrete mixers and tower cranes became more widely used from the 1950s, and tubular steel scaffolding replaced rope-lashed timber scaffolds. Trends in the development of construction machinery during this period included the development of smaller and larger machines, higher performance, longer life, less maintenance and repair and better transportability.

By the 1970s, the relevance and appropriateness of Fordism was increasingly being questioned. Changes in the wider and market environments demanded new directions in construction's products and processes. A significant factor was the oil crisis in 1973, the rapid increase in the cost of energy, and the need for greater energy conservation. This required that all buildings became more resistant to the passage of heat. In office and commercial buildings this also manifested itself with a reconsideration of the attractions of natural lighting and in some cases natural ventilation. It again opened up the arguments for and against the deep plan factory with an artificial environment, or the naturally lit plan types. It also reflected a shift of emphasis in the sociology of factory and office work, the growing reaction against the formality and rigidity of Fordism and a call for the adoption of less formal and more flexible approaches. This was also reflected in a desire to return to smaller factories due to bad labour relations experiences in large, centralised centres of production and the return to public housing on a more human scale. During this period, when interest rates were in excess of 20%, it was important to clients that projects were completed quickly to reduce charges on borrowings and provide quicker returns on investment through earlier income flows. The industry responded by developing 'fast-track' construction, which primarily consisted of a combination of off-site prefabrication and overlapping the design and construction stages as a means of reducing overall project times.

Despite the major changes and innovations in construction's products as outlined above, there is considerable evidence to indicate, however, that throughout the Fordist era, the British construction industry lagged behind those of the USA and some other European countries. The uptake of new materials, such as steel, was adopted much more quickly in the USA and the use of concrete and prestressed concrete advanced more quickly in France. Also, the adoption of mechanisation was slow in comparison with other countries. Morton (2000) points out that in 1959, the Building Research Station (BRS) listed the processes that had then become mechanised or partly mechanised including joinery, plumbing (through prefabrication), the vertical and horizontal movement of materials and components, the excavation of trenches and pits and the mixing of concrete. The BRS argued that the UK was slow to adopt such mechanisation, giving the example of the tower crane, which was invented in France in 1858 but only imported into the UK in 1950. A survey undertaken in the mid-1970s did not show any significant changes in this list, which suggests that construction is slow to change.

The above outline of changes in construction's products shows that although construction did respond to the changes in its market and wider environments during the Fordist era, there were also major barriers to innovation arising from what Bowley (1966) called its outworn pattern of organisation. The characteristics of construction's operating system and its attempts to address its weaknesses are outlined in the following section.

The impact of Fordism on construction's operating system

The objective of this section is to review how the industry has changed its operating system during the Fordist era. The type of contracting strategy and procurement approach adopted by clients and their development advisers largely determines the project environment and the operating system within which all organisations in the process operate. Hence, it can be argued that the industry's development of alternative building procurement systems and contracts can be seen as an explicit representation of the extent to which Fordist thinking and principles permeated its organisational system.

Although there is a substantial body of literature covering the methods of construction procurement, there is no consistent approach to classifying each method, their many derivatives and associated contracts (Cox and Townsend, 1998). Franks (1998) identifies four main procurement systems and six routes or methods available during this period. These are 'traditional', 'management contracting', 'construction management', 'package deal turnkey', 'design and build', and 'British Property Federation'. Up until 1984, surveys of the use of the various procurement methods did not exist. Until 1984, around the end of the Fordist era, when the first RICS survey was conducted, the majority of building work, probably more than 80%, was carried out using the traditional single-stage tender system. The different procurement approaches associated with Fordism are discussed below and summarised in Table 2.1.

For centuries prior to the Industrial Revolution construction was organised on the basis of Craft Guilds. There were two levels of membership of the Guilds: the masters and journeymen. The masters owned their own businesses and employed journeymen and apprentices. The journeymen comprised those who had served apprenticeships but were unable to acquire their own business because of insufficient capital.

Table 2.1. Main procurement systems associated with Fordism

Procurement method	Key characteristics	Impact on the management of innovation
Traditional single-stage	To raise output through industrialisation Designer (often architect) led Specialisation through separation of design and construction Cost-oriented approach through competitive, fixed-price tendering Clear lines of accountability	Allows architect to focus on the client's needs Drawings and BoQs provide common basis for competitive tendering and valuation of variations Design team retains control over design intentions which can result in high quality and functionality of the completed building Some degree of cost certainty at the start of construction Flexibility for design changes but with considerable cost and other implications Independent advice available on most aspects of the process with independent verification of performance Fragmentation, hierarchy and division of work with different professions and subcontracting Emphasis on contracts fosters opportunistic relationships which can easily become adversarial instead of collaborative Architect nominates the design team and has to control the team members but has no contractual liability for their performance and little project management training or skills Slow and convoluted decision-making Low buildability as there is no opportunity for contractor and subcontractors to contribute to design Many technological and business interfaces to be managed Encourages outsourcing and short-term relationships between contractor and subcontractors Long total project times as design and construct are undertaken sequentially

(continued)

Table 2.1. (contd.)

Procurement method	Key characteristics	Impact on the management of innovation
Two-stage tendering	To improve integration between design and construction through an earlier involvement of main contractors	Allows early testing of market to establish price levels Opportunity for the main contractor to add value and manage risks at the design stage, particularly in relation to buildability and cost determination before the design is complete Potential for greater integration but countered by continuation of outsourcing strategies by main contractors leading to more fragmentation More client involvement needed and possible
Fast-track construction: management contracting	To reduce project time by overlapping design and construction stages and providing an early start to construction work Used where the project is large or complex and where there is a need for an early start and completion An organisationally complex approach involving the management and coordination of a large number of contractors	Time can be saved by overlapping design and construction Allows flexibility for change and last-minute decisions Cost savings possible through better control of design changes, improved buildability and planning of design and construction into packages for phased tendering, and keener prices due to increased competition on each package The managing contractor has direct contact with the client as in the case of design consultants Potential for coordination and integration Scope for value management and engineering Complex contractual relations and assignment of liability Conducive to 'buck-passing' Potential for more conflict between participants as roles change Client exposed to greater risk because of incomplete information and potential failure of specialist contractors Potential for misunderstandings at the interfaces between packages and works contractors

Procurement method	Key characteristics	Impact on the management of innovation
		May cost more but can be offset by earlier completion and hand-over Overlapping of roles and administration
Fast-track construction: construction management	The construction manager acts as the client's representative and has authority to make decisions affecting the designers and the works contractors Requires experienced client with necessary commitment Contractor acts as an impartial client's agent controlling all aspects of the project	Depends on the increasing professionalism of the main contractors who are appointed early to undertake the management of the works for a fee Offers considerable flexibility Allows the full and continuous involvement of the client who is the ultimate decision maker and arbiter between the parties and on issues in relation to design, functionality, cost and time Client can have considerable say in appointment of construction contractors Scope for value management and engineering Direct relationship between client and construction contractors can reduce costs and add value – particularly in serial contracting Conducive to teamwork and collaboration Client carries the total risk in event of failure as there is no intermediary contractor Can be daunting for inexperienced clients
Package deal, turnkey, design and build	To create a single-point responsibility, low-risk and quicker development for the client	Needs a clear and detailed brief otherwise competing schemes may not meet client's requirements Solutions not always fully thought out or focused on the end users needs After selection it gives single point responsibility for the design and construction process Integration of design and construction High levels of buildability through the early involvement of the contractor

(continued)

Table 2.1. (contd.)

Procurement method	Key characteristics	Impact on the management of innovation
		Encourages the use of off-the-shelf solutions Competition on product as well as price Can speed up commencement times and provide shorter development times Lack of independent advice Some evidence of trust between clients and main contractors Potential to rationalise and align key downstream participants and processes Potential for rationalisation of products and processes particularly for standardised building types Often high levels of competition for packages despite potential for longer-term relationships and preferred supplier status Variable design quality and functionality because design and build firms can lack the breadth of experience in traditionally procured teams Variations and changes to design can be expensive, can be only with the contractor's consent and are difficult to price because of the lack of standard rates Tender process expensive and wasteful of bidders' time and resources Comparison of bids can be complex
Nomination	Recognises the contribution of key subcontractors in order to exploit their technical knowledge and skills	Gives clients and their advisers more control over the selection of key subcontractors Favoured by subcontractors because it gives them contractual protection and allows the development of long-term relationships with regular clients Complex contractual relations Reduction in competition between subcontractors without robust performance measurement Restricted choice of building elements, components and materials

There were two normal places of work for the craftsmen: shops and, of course, construction sites. The masters became rich and soon acted as merchants, leaving the craft work to the journeymen and becoming employers of labour. In terms of construction's operating system, clients approached masters who would design the building and employ journeymen and apprentices to undertake the work which they supervised as shown in Fig. 2.2.

Within this operating system, the design of buildings was often undertaken by these masters until, in the fifteenth century, a new breed of master/journeyman took on the role of principal master. This was the earliest form of architect. By the seventeenth century, architectural designing was well established and by the early years of the nineteenth century, an owner requiring a building would commission an architect – who by then was normally a gentleman educated in classical design – to produce drawings. The structure of this operating system is illustrated in Fig. 2.3.

Once a design was agreed, the architect would engage a master craftsman for each trade, on behalf of the owner, to undertake the work. The craftsmen's work was coordinated by the architect's overall design with details being worked out on site through direct discussion between

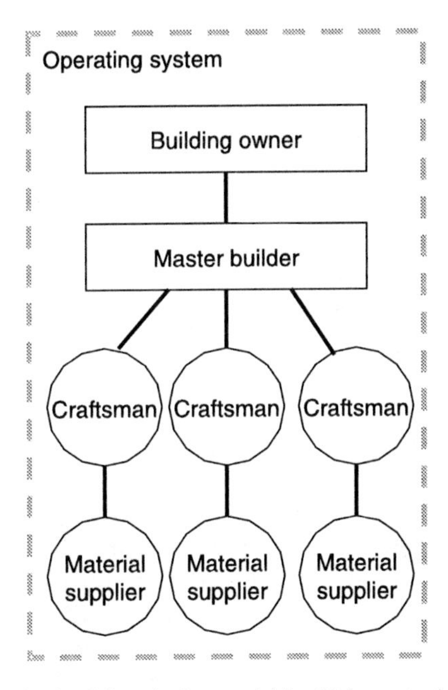

Fig. 2.2. Structure of the building industry 14th–15th centuries

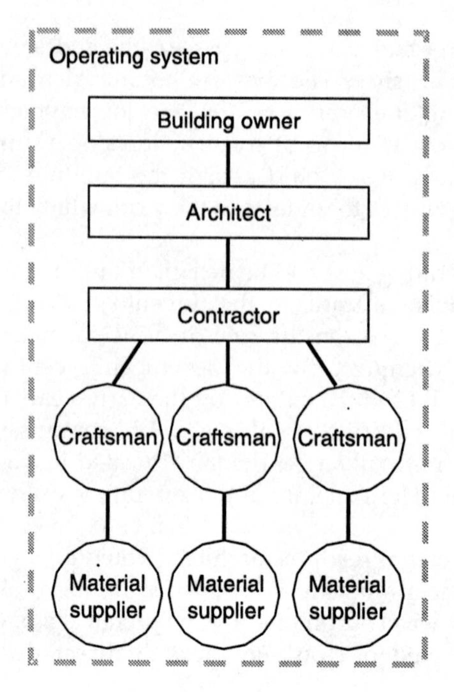

Fig. 2.3. Structure of the building industry 17th–19th centuries

master craftsmen and the architect. This can be seen as mirroring the principle of horizontal and vertical division of labour being followed in the wider environment and other sectors of the economy with the rise of industrialisation and the factory system. The price for the craftsmen's work was agreed with the architect acting as owner's agent. This was a straightforward process as there was a standard time and price for each type of work in each area of the country. However, the craftsmen felt disadvantaged in their dealings with the better-educated and socially superior architects and began to employ measurers to look after their interests. These forerunners of quantity surveyors measured work after it was completed and agreed a fair price with the architect on behalf of the craftsmen.

With the advent of the Industrial Revolution, a new approach to the system for the design and construction of buildings and other structures began to emerge, and was developed through the nineteenth and twentieth centuries. This later became known as the 'traditional' approach. This 'traditional' system or approach was needed to respond to increased demand for construction products and services and accommodate changes in market forces and technical innovations outlined earlier in this chapter. As well as responding to the increased and changing demands arising from industrialisation and urbanisation and taking advantage of technological innovation in other sectors of the economy, it also responded to the new system of organising work. This new system included the principles of the horizontal and vertical division of labour, dividing of tasks between mental and manual labour, and high levels of specialisation.

Early in the nineteenth century, the government adopted a general contractor approach in which one of the master craftsmen, often a bricklayer, took responsibility for coordinating the work of all the trades and employing them as subcontractors. The general contractors were appointed on the basis of lump sum tenders for complete buildings. So that they could provide accurate lump sum prices when competing for new work, measurers had to learn how to measure from architect's drawings. In addition, as construction's clients became more dominated by entrepreneurial and often ruthless businessmen, the industry faced demands for greater speed and economy. These developments necessitated changes in working practices. This modified version of the former approach to procurement involving architect, quantity surveyor and general contractor is shown in Fig. 2.4. It forms the core elements of many of today's approaches to the procurement of buildings and other structures and its operating system.

As well as demanding alternative structures, the emergence of larger and more complex projects during the Industrial Revolution also gave rise to larger contractor organisations. Railway construction in the UK, which began in the 1830s, was a major factor in the emergence of large national, and later international contractors. The first British railway contractors were mainly self-taught practical men who learned by experience. When the railway building programme declined at the end of the 1840s, British contractors sought and won contracts abroad on the basis of their experience. Some of these contractors, such as Thomas Brassey, Samuel Morton Peto and Edward Bettes were, at the peak of their activities, providing work to tens of thousands of people. Around the turn of the century, Pearson, another UK contractor, built up a major international construction company in the UK and when he accepted a 1 million acre oil drilling concession in Mexico as part payment for his services, he showed an early example of diversification, which later became more common among large contracting firms (Morton, 2000). The appearance of international contractors at this time can be seen as a significant step in the internationalisation of construction.

The key characteristics of the traditional or general contracting approach to procurement include the separation of the design and construction processes with the employment of separate consultants to

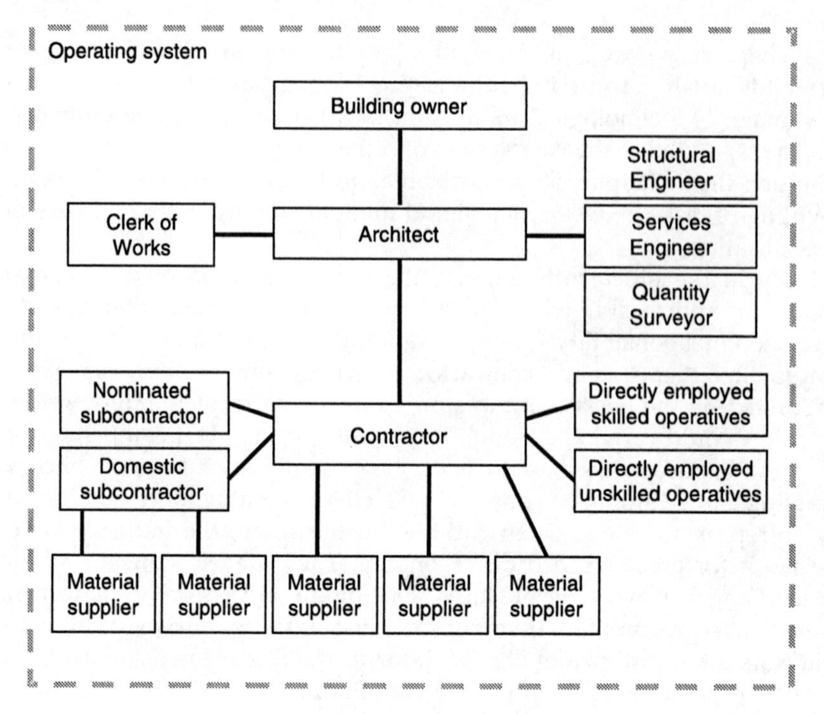

Fig. 2.4. Structure of the building industry 19th–20th centuries

design the building or structure, a contractor to take charge of the construction process, and subcontractors to undertake the work. This means that there is little integration across the design and construction interface and the design and construction processes are seen as sequential and independent. The contractor is selected through a competitive tendering process on the basis of a fixed price bid. The contractor's input to the design process is minimal (often nil), and most (often all) of the work is subcontracted to subcontractors, again on a fixed price basis following a competitive tendering process. In contractual terms, the client makes separate contracts with each of the building professions and the main contractor. In turn, each of the subcontractors then makes a separate contract with the main contractor. The contract is then awarded to the successful contractor on the basis of a lump sum. This is not usually fixed as the contract contains many clauses and elements that may cause the final cost to vary from the tendered cost.

As the range and complexity of building products, technologies and materials increased during the Fordist period, it became more difficult for one person to solve all the problems at each stage in the process. In this situation, increased specialisation became inevitable. What was not logically inevitable, however, was that each specialisation in construction was segregated into separate compartments in terms of both training and the use of knowledge. Typically, where an architect was employed the relationship between the architect and the other participants tended to be that of employer to employed. The architect designed the building and if he (invariably the architect was a man in those days) was not able to design the structure to support it, he passed his problem to an engineer. The possibility that an engineer's knowledge of structures might have suggested a more economical, efficient or even more interesting or beautiful building was not normally considered. Similarly, by the time the design, complete with the engineer's contribution was passed to the contractor, there was no opportunity for the builder's knowledge to influence it. Bowley (1966) argues that it is difficult to see how any system more wasteful of technical knowledge, intellectual ability and practical and organising experience could have been invented.

The separation of the designer from the erection of the building meant that he had no means of determining how particular items of the design or layouts affected cost. This gap was not filled by the quantity surveyor whose main function remained that of developing a form of bill of quantities, which would enable contractors to estimate their total tender prices. The other task of the quantity surveyor was to measure variations and other work not included in the original specifications. The bill of quantities was not designed to enable comparative costs between alternative forms of construction. Furthermore, the contractor's margins for overheads and profit were not included. Although quantity surveyors could have advised the designer on costs, they lacked the scientific basis and expertise to undertake such a detailed comparative analysis.

The traditional form of contracting had some advantages for clients in that they could exercise direct control over, or intervene at, various stages of the construction process, which helps to explain why it remains the principal form of procurement in the UK. Clients can take advice from different sources and, within limits, change the design and specification of the building as the construction work proceeds. There are also a variety of interests that police the activities and performance of the various agencies involved in the project. Yet, despite these advantages to the client, many of the problems and deficiencies associated with the construction industry are said to stem from the traditional contracting system. For example, the division of the project into separate contracts with divided decision-making makes it difficult, if not impossible, to build a unified project team of managers, designers and workers, or to provide smooth progress of the project and continuity of work for the participants. Other problems arise from the divisions between contractual obligations and functional roles, which result in some agencies involved in the project having powers but little or no ultimate responsibility.

Difficulties predicting tender prices under intense price competition also generate pressures for contractors to realise profit by maximising claims against clients and subcontractors. Improving the efficiency of the construction process is usually subordinated to the aims of generating a profit, maintaining a portfolio of ongoing projects, avoiding or shedding risk, and maintaining flexibility of financial assets for investment either within or outside construction.

A further weakness of the system lies in the short tendering periods allowed to specialist and trade subcontractors – none more so than the M&E specialists. M&E installations are normally one of the largest subcontracts on a project and the subcontractor often needs to sublet sections of its work to other specialist contractors. This means that the two-week tender period often allowed by some main contractors seems quite unreasonable. As well as the limited time available for tendering, an M&E contractor might spend, on average, 0.5% of its overhead costs to tender for the project. If it only wins one in six projects, the M&E contractor would have to recover 3% of overheads to break even. With continuing pressures on margins in a price-competitive environment, it is understandable why main and specialist contractors are keen to move away from traditional tendering methods.

A number of attempts have been made to reshape this traditional procurement approach to address its limitations. For instance, the two-stage tendering process was introduced in the 1960s to improve the integration of design and construction and involve the main contractor earlier in the process. In two-stage tendering the design team establishes a notional bill of quantities when the design has reached an appropriate stage in its development. Selected main contractors are then asked to tender rates against approximate quantities contained in the bill. They are also requested to provide proposals for the management of the construction and any suggested design changes to improve buildability. The successful main contractor then becomes a full member of the project team. However, the potential of the two-stage tendering process for downstream integration was not generally extended beyond the main contractor. The opportunity for a similar involvement of specialist and trade subcontractors was to some extent countered by a combination of the continuing emphasis on price competition by clients and their advisers in selecting the main contractor, and the main contractors' response based on a strategy of 'financial optimisation'. This encouraged further downstream fragmentation from vertical disintegration of main contractors and the wider use of outsourcing and subcontracting.

There have been some more specific attempts to recognise the growing contribution of specialist subcontractors and their supply systems, further unlock their potential, and involve them in projects alongside the main contractor. One such approach is 'nomination' which emerged during the early 1900s as a way of securing early specialist contractor involvement to deal with the increasing technological sophistication of buildings. The primary motivation behind the adoption of nomination was the need to exploit the specialist subcontractors' skills, particularly their design skills, before the appointment of the main contractor, and to allow for the preparation of building components with long lead-times so that they were ready for installation in the building at the appropriate time. A further motivation was that nomination also gave the client and their advisers control over the selection process for key specialist trade subcontractors on the basis of quality as well as cost and time (Franks, 1998). Subcontractors favoured nomination because it gave them contractual protection and allowed the development of long-term relationships with regular clients based on mutual advantage. However, main contractors were resistant to nomination because it limited their scope to select subcontractors and they resented the subcontractors' preferential relationship with the client, which was seen as undermining their power and influence.

Although this modification had considerable potential for the involvement of specialist subcontractors, its use declined because of a number of problems. The three-way contractual relationship between client, contractor and subcontractor, particularly where a nominated specialist contractor fails to perform properly or becomes insolvent, was very complex. Also, clients were concerned by the reduction in competition between specialist subcontractors when lists of preferred suppliers become small. The main contractors were concerned by the reduction of their influence on nominated specialists because of their closer relationship with clients. The main contractor's motivation to control the expenditure on nominated subcontract packages was reduced in comparison with priced work (CIRIA, 1998).

Further alternatives to the traditional approach emerged following the oil crisis of the early 1970s when the time taken to complete construction works became more important to clients than its cost. During a period when interest rates were in excess of 20%, it was important to clients that projects were completed quickly to reduce charges on borrowings and provide quicker returns on investment through earlier income flows. The industry responded by developing 'fast-track' construction, which primarily consisted of overlapping the design and construction stages as a means of reducing overall project times. Such 'fast-track' construction was achieved through the adoption of construction management or management contracting procurement methods. Typical systems are shown in Figs 2.5 and 2.6.

There are many variations of the two approaches detailed in Figs 2.5 and 2.6 and precise arrangements and definitions vary considerably.

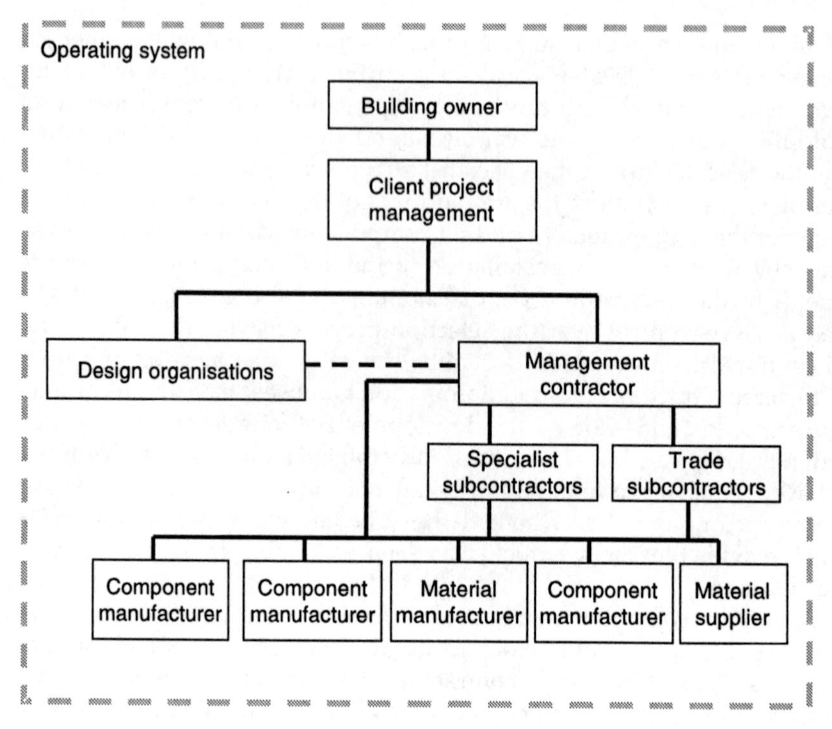

Fig. 2.5. Management contracting operating system

Their most significant common feature is that the management contractor or construction manager undertakes the management of the works for a fee, having in effect the same relationship with the client as the architect or any other consultant. The main difference between construction management and management contracting is the relationship of the client to the subcontractors. In the construction management system the client contracts directly with subcontractors and there is no contractual relationship between the main contractor and the subcontractors. The construction manager acts as the client's representative and has authority to make decisions affecting the designers and the subcontractors. The subcontractors are appointed on a competitive and firm-price basis. The client enters into contracts with numerous subcontractors rather than a single contract with the general contractor as in the case of the traditional system. When, as in this case, contracts are made directly between client and specialist contractor, contract conditions can be adopted which are more appropriate to the nature of the works to be undertaken. Also, the lines of communication between clients and specialist and trade subcontractors can be shorter than with the tradi-

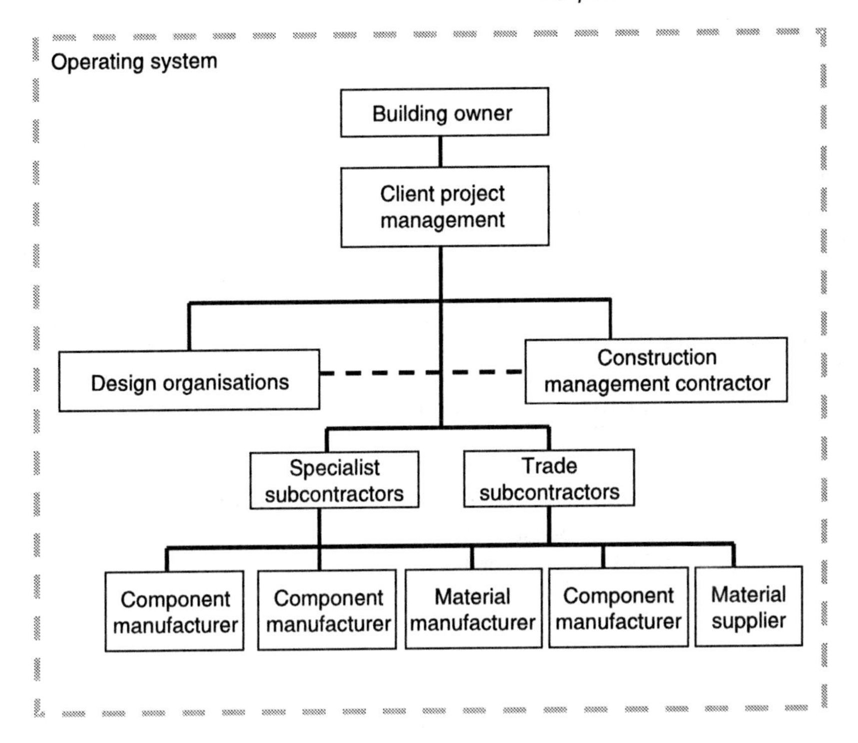

Fig. 2.6. Construction management operating system

tional system, leading to quicker decision-making by the client in relation to unforeseen site problems, and by the contractors in relation to changes required by the client.

The main advantage of both these approaches is that the management contractor and construction manager are appointed much earlier than is possible with the traditional system, allowing their construction knowledge and expertise to be placed alongside that of the design team. Decisions regarding the appointment of the specialist and trade subcontractors can be made jointly by the designers and construction managers offering more opportunities for their selection on a wider range of performance criteria rather than just price. However, clients, particularly the irregular and infrequent users of construction's products and services, consider the increased level of risk they are expected to carry to be unacceptable. This means that these approaches have proved to be more appropriate for the regular and very experienced construction client.

As a response to many clients' demands for single-point responsibility, low-risk development, known total financial commitment and quicker procurement, the industry has developed single source systems. This is a group of procurement systems that enables clients to employ one firm to take responsibility for the complete delivery of the project. There are a number of variations including 'package deal', 'turnkey' 'design and build', 'design-build-finance', and 'design-build-finance-operate'. A typical design and build operating system is shown in Fig. 2.7.

The principle underpinning all these systems is that the client contracts with one organisation, usually the main contractor, for the whole of the design and construction process. The main differences between each of the variations is the scope of the responsibility allocated to the package dealer. As responsibilities are not split between designer and builder, the client is not involved with separate parties in the event of a building failure. The risk to the client is reduced, although clearly the cost of accepting that risk will be incorporated in the tender produced by the package dealer. Using this method, means the client's total financial commitment is known early in the project, provided that changes are not introduced during the course of the works. Competition or negotiation may be used to select the package dealer and its use can open up competition to a wider range of performance criteria rather than price alone.

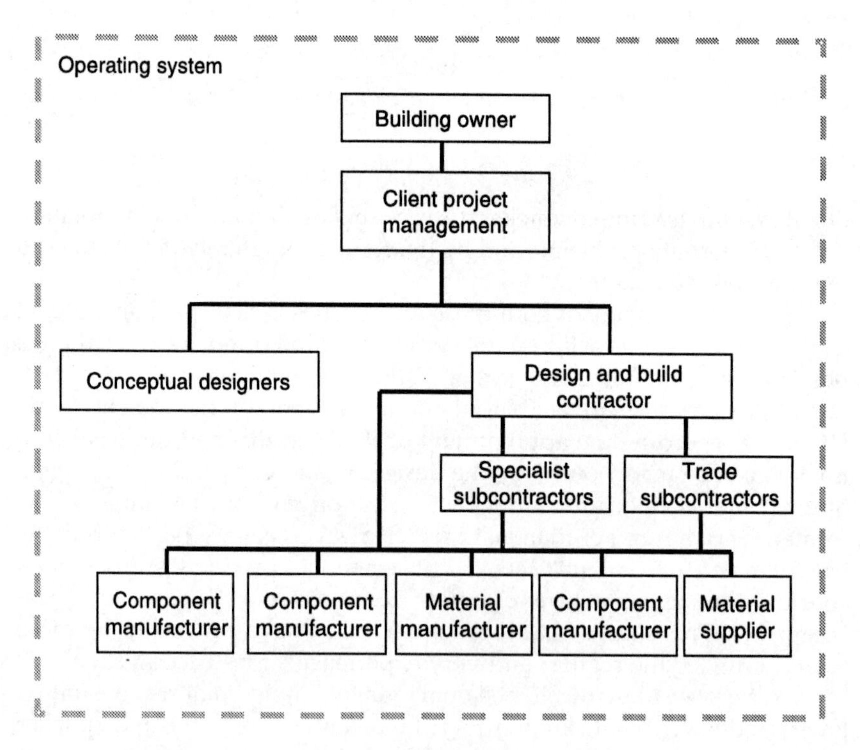

Fig. 2.7. Design and build operating system

As the client has direct contact with the design and build contractor, the lines of communication are shortened which enables the contractor to respond and to adapt more promptly to the client's needs. Given the more integrated nature of the service provided, the specialist and trade subcontractors and their suppliers need to be appointed early in the process. This means that the method is best suited to projects with low levels of uncertainty as in the case of offices, housing, out-of-town retail centres and distribution warehouses. Indeed, in such circumstances, many package dealers provide a comprehensive package of services including raising the finance, finding and purchasing a site, obtaining planning permission and building regulations approval. As it is best suited to more standardised products, there are more opportunities for the package dealer to integrate specialist and trade subcontractors and their suppliers into the process. A number of package dealers have gone further in rationalising the supply side by using a proprietary building system or modular building form. This helped in shifting construction activities from the site to the factory and reducing the time needed for design and for the approval of the building components. However, in spite of its potential for more collaboration there can still be high levels of downstream fragmentation and intense price competition for trade packages.

It needs to be borne in mind, however, that although the above review, summarised in Table 2.1, defines and classifies procurement approaches, there are no generally accepted formal structures or agreement on the terms. Also, many of the systems share many characteristics. This, Hibberd (1991) argues, implies that the terms used and definitions given imply a degree of scientific rigour which simply does not exist. However, it is clear from the above review of procurement approaches, that construction's structure and procurement processes were significantly influenced by the traditional contracting system that emerged during the early years of the nineteenth century. It is also clear that from the 1960s onwards construction has made considerable attempts to change its operating system and address its hierarchical structure and class relations. Despite these efforts, for much of the Fordist period construction's operating system and social structure remained dominated by people lacking moral scruples with elitist views (Ball, 1988). This means that relationships remained essentially opportunistic and adversarial. Contractors continued to gain the maximum advantage from their pivotal merchant position within the operating system – purchasing inputs for production from subcontractors and suppliers and selling on the final product to clients. The architectural and engineering design professions retained their conservatism and detachment from contractors. It is only with the decline of the traditional contracting system based on Fordist thinking and principles from the mid 1980s onwards that we see the emergence of a new social structure with greater emphasis being placed on more equitable and synergistic relationships between the parties, greater integration, longer-term relationships, earlier involvement of key specialist and trade subcontractors and more engagement with supply systems. These more recent attempts to increase collaboration and integrate the design and construction team are described and analysed in Chapter 3.

Another significant aspect of the operating system is the management style of the main contractors. Writing at the height of the Fordist era, Bayley (1973) observed that the construction industry had an above average number of enterprises where the major shareholder is 'himself' the controlling director. In such cases, the leadership as well as the tone of the organisation emanates from 'him'. He argued that it is not unusual in such circumstances to find the authoritarian or sometimes paternalistic style of management. The authoritarian manager takes the 'direct' and 'command' aspects of management as almost a divine right. Labour is seen as a commodity to be hired in similar manner to plant and materials. The formal representation of the labour force outlined in the following section was not seen as a potentially constructive component of business. This manifested itself in an inbuilt sense of hostility towards unionism and the unions' increasing power. Bayley identified a growing trend towards the employment of professional managers in the larger national companies and a decline in the influence of shareholder-directors. These professional managers were more receptive to modern management with a more conciliatory approach to labour, which placed the contribution of the 'heads' of labour alongside its 'hands'. However, Bayley (1973, p. 9) identified that

> *the climate of industrial relations within the industry in many respects continues to be influenced by the old authoritarian style which is coming under increasing pressure to change its approach.*

The emergence of the trade unions

The late eighteenth and early nineteenth centuries saw increasing separation between 'heads' and 'hands' within its operating system as it became increasingly influenced by Fordism, with its emphasis on the division and specialisation of work. As the divisions between contractors and construction craftsmen became more distinct as a result of the introduction of the contracting system, groups of workers began to form local clubs and friendly societies. These were set up to provide mutual support and, like the Guilds from which they were descended, took on the responsibilities for helping unemployed members, giving financial support to widows and children as well as seeking to maintain craft standards. In the early years of the nineteenth century, trade unionism in general was weak and the state active in limiting its power (Pelling, 1976). The long period in the development of the trade unions is characterised by four main phases during the Fordist period. The early years of industrialisation was characterised by fragmentation and temporary unions and intense struggles. These were superseded in the 1860s by a formalisation of industrial relations in the industry. From the early years of the twentieth century to the 1940s, building unionism was reshaped through the amalgamations of craft unions and the growth of union membership outside the traditional craft boundaries (Ball, 1988). From the 1950s onwards, craft-based unionism declined. At a general level, the development of trade unionism in construction has at certain times included strikes and lockouts but also repeated attempts to find common cause with employers and establish national forums for negotiation on working conditions and rates of pay (Morton, 2002).

The early strikes were instantaneous and involved work groups rather than union members. Although strikes were often over local issues, often they would reflect wider developments within the industry such as rates of pay and attempts to increase productivity. Employment of subcontract, particularly non-union labour, was a constant source of friction, as was the number of apprentices with employers wanting more to drive down rates of pay and the unions wanting fewer apprentices to maintain levels of remuneration. As well as disputes over working conditions, hours of work and piecework, there were also conflicts in relation to the organisation and control of the process with respect to such matters as subcontracting, piecework and 'driving' foremen (Ball, 1988). Indeed, one of the early disputes in construction was started by stonemasons over the aggressive behaviour of the foreman on the construction of the Houses of Parliament in 1841.

The switch from day to hourly payment in the industry during the 1860s was resisted by the workforce as it weakened their control over production, leading to long working hours in the summer and unemployment in winter (Price, 1980). Ball (1988) argues that the point at issue was not just levels of pay and working conditions but also the power of management or workers to change them. The first major conflict that involved unions from all trades together with labourers started with a demand for a nine-hour day by carpenters and joiners in London in 1859. Ultimately, the union lost over the nine-hour day and the employers brought in hourly instead of weekly wages and extended the use of piecework – both bitterly opposed by the unions. The last two decades of the nineteenth century were a period of boom, unemployment was low and the unions grew stronger. As union power grew, so the employers saw it in their interest to form associations to fight jointly against what they saw as the unions' unreasonable demands for workers. The struggle continued with bitterness and ferocity but for thousands of workers life was still hard with no job security, low pay (especially in periods of low building demand) and no insurance or compensation for injury.

Although disputes and the harshness of working conditions continued, by the turn of the century there was a growing consensus that the way forward was through negotiation (Morton, 2002). This corresponds to the emergence of the School of Human Relations Approach, which started challenging the authoritarian style of management which characterised the traditional approach. After a few false starts, the machinery for negotiation was established in the National Joint Council for the Building Industry in 1926. The results of negotiations were codified in the Working Rule Agreement, which exists today, although in modified form.

Despite the growing emphasis on negotiations from the early years of the twentieth century, from the mid-1960s the relationship between main contractors and their workforces again deteriorated. This can be seen as part of a wider movement of protestation against the Fordism philosophy, which was increasingly taking place around the developed world. With the continuing emphasis on this traditional approach, pay was tied to work rates through bonuses and overtime. In addition, the adoption of industrialised building systems resulted in the construction process being divided up into tasks that did not require traditional craft skills. These developments met with significant trade union resistance resulting in a number of site-level disputes.

One method used by contractors to circumvent poor labour relations was to increase the use of subcontracting. This approach either transferred these labour problems on to another company or, through the growth of labour-only self-employment, removed the role of trade unions altogether. However, in retrospect, subcontracting transformed rather than removed the labour difficulties faced by main contractors who gradually began to lose control over labour. Consequently, site labour relations became more inter-organisational, contractual and adversarial.

The emergence of the professions

The early development of Fordism in construction helped to create the rigid formalisation of the professions and their separation from contractors. Prior to the Industrial Revolution, there was no distinction between an architect and a master builder. However, some architects who understood classical and Gothic precedents began to acquire a higher status by working for the aristocracy, church or government. In 1791, an exclusive Architects' Club was set up by four men who were well-known as builders as well as architects but it was the British Institute of Architects (founded originally as the Architectural Society in 1831) that marked the real move towards the modern profession, receiving the designation 'Royal' from Queen Victoria in 1866.

In 1887, the RIBA made a decision that fundamentally affected the way in which construction's operating system was conducted. They sought to maintain the independence of their position by restricting the conditions under which their services were sold when they prohibited any member of the RIBA from holding a profit-making position in a capitalist enterprise. Morton (2002) argues that this had two major consequences for the operating system process. First, builders did not have their sons trained as architects and, second, architects could not run construction companies. Although the motivation behind this decision was understandable in that it was aimed at protecting the interests of the client, it drove a wedge between architecture and the management of construction. This major division between design and construction was not rectified until 1981, but by then the separation of the roles had become deeply embedded in construction's operating system and procurement approaches.

Architects found it increasingly difficult to deal with the proliferation of new techniques and materials brought about by the Industrial Revolution. This gave rise to the emergence of engineering specialisms and further fragmentation of the professional advice available to clients. The Institution of Civil Engineers was founded in 1818 as a study association having no codes of conduct. Their status grew as their importance was recognised with the increasing need for canals, railways, bridges and docks. As a response to the increasing sophistication of building structures, the Institute of Structural Engineers was established in 1922 and was granted a Royal Charter in 1934. The Chartered Institute of Building Services Engineers was created only in 1976. It was an amalgamation of two much older organisations, the Institution of Heating and Ventilating Engineers founded in 1897, and the Illuminating Engineering Society, founded in 1909.

The construction of the new Houses of Parliament in the 1830s saw quantity surveyors being used in their main specialist function, as this was one of the first buildings to have been approved on the basis of detailed drawings and a bill of quantities. This set the future role for the quantity surveyor – estimating from the detailed drawings and specification, the quantity of labour and materials required for each part of the building. The bill of quantities produced became the means against which contractors could tender and hence prepare a detailed and accurate bid all on the same basis. Although quantity surveying came to be seen as a fundamental requirement in the UK operating system, it did not develop separately elsewhere and there are good reasons to suggest that the accurate methods of estimating and cost prediction it implied were never economically sound. The differences and tensions between client and contractor are also reflected in the surveying profession. Their independent advisory role to clients is quite different from the roles they undertake as contractors' quantity surveyors. Ball (1988) argues that the physical division between the two types of surveyor is very much a product of their place in the social division of labour in the UK construction industry.

The idea of professional construction management as a profession is relatively recent. However, a number of builders did emulate architects and engineers and formed associations – usually to protect themselves against the growing trade union movement. The earliest organisation was probably the Builders' Society formed in 1834. It eventually came to perform the role of a professional institution, evolving into the Institute of Building and achieving chartered status in 1980.

This increasing level of specialisation and the emergence of a wide range of professions are often explained and justified by the increasing level of task complexity in construction. However, this has led to a greater division of tasks creating a hierarchical and fragmented structure as the professionals are involved, first, in the design of buildings and second in the measurement and costing. These professionals intervene in, and indeed shape, the relationship between client and contractor through the specification of the procurement method and form of the contract. Through the contract, the work of the contractor can be controlled by the professionals in the interests of the client. In turn, the professionals can use the contract to agree claims for extra work by contractors. This means that the role of the professionals – particularly architects and surveyors – is ambiguous. It can be argued that this ambiguity in the operating system has constituted a major barrier to progress and to innovation in the industry.

Supply systems

Prior to the Industrial Revolution, construction's supply systems provided the basic building materials to the main trades – bricklaying, stonemasonry, carpentry, roof tiling and slating and plastering. Most buildings were constructed and fitted out with easily available, local materials. Building styles and forms of construction tended to reflect the availability of indigenous materials and the skills to make them. Much of the preparation of these materials was undertaken in the master builders' yards and on site. The exceptions were prestigious buildings, which justified the expense and effort of transporting heavy, bulky materials over longer distances using animal-powered transport over poor roads.

During the nineteenth century, improvements in transport including better roads and the development of canals, and later railroads, made it economically viable to transport building materials over long distances from their point of extraction or manufacture. In the second half of the nineteenth century, modern factories were established to produce building materials including cement, lime, gypsum, steel, bricks and roofing tiles. Manufacturers produced catalogues describing and promoting their products so that designers could choose and specify their products accurately. By the beginning of the twentieth century, a new industrial sector producing building equipment including elevators, boilers, radiators, pipes and sanitary appliances was emerging. During this period, a wide range of new building materials and components were developed as were methods of using traditional materials in new ways. Most of these innovations and improvements often took place inside these growing supply systems and were later assimilated by designers and contractors. These innovations were often stimulated by factors outside the industry such as social and market changes.

Specific events also affected demand for construction materials. The reconstruction work after the Second World War further promoted the growth of production in the building materials industry and encouraged the adoption of Fordist approaches by the producers of building materials. For example, in the UK, the growth in cement production grew from 8.5 million to 12.0 million tons between 1948 and 1957. World steel production amounted to 150 million tons in 1950 and 600 million tons in 1970. However, this growth in production slowed again during the uncertainties of the 1970s, particularly following the major economic disruption to the world economy following the oil crisis of 1973.

As well as demands for new materials and components, there were new and increasing demands for construction plant and equipment. All construction involves carrying, lifting and placing materials, either in an unformed or preformed state. Materials often have to be cut, shaped and jointed. In recent years it has also increasingly involved operations including stressing or jacking or assisting in the construction process through the pumping of liquids and semi-liquid materials. Substructure work also involves operations such as excavation, piling and pumping away, excluding or controlling groundwater. In the later nineteenth and twentieth centuries, more diverse structural systems called for a greater variety of temporary support systems.

The period between the 1950s and 1970s saw particular advances in the technologies used in construction processes. Mobile concrete mixers and tower cranes became more widely used and tubular steel scaffolding came into universal use. Inventions in other areas meant that, increasingly, construction machinery could be controlled by mechanical, hydraulic or pneumatic means or a combination of each, rather than the traditional rope-operated machines. Trends in the development of construction machinery during this period include smaller and larger machines. It also resulted in higher performance, longer life, less maintenance and repair, better transportability, reduced fumes and noise, less vibration and operator impact, and improved safety features for operators. The manufacture of such construction plant was dominated by Germany, the USA and Japan.

Gann (2000) argues that many of the changes in the process aimed at improving performance have taken place off site in the supply systems where materials and component suppliers and plant and equipment manufacturers are better placed to make longer-term commitments to research and product development. Increasingly, value is being added downstream in the supply chains by component and material manufacturers but has also led to a further de-skilling of the workforce on site. Also, rapid and ill-considered applications of new materials, components and techniques led to mistakes and products and processes less robust than traditional solutions. Managing these issues and deriving maximum competitive advantage from the supply systems through the adoption of the principles of supply chain management, is investigated in Chapter 7.

Changes in the regulatory and institutional framework

The increasing control over construction activities began when the excesses of the Industrial Revolution became more apparent and with the emergence of a regulatory framework. As construction affects important areas of public concern such as health and safety, public authorities of all types have sought to regulate various aspects of construction. In the early years, industrialists and speculators determined the way land was used, or indeed abused. Development was often haphazard with the main criterion being how to accommodate employees as closely and cheaply as possible to their place of work. Little attention was paid to layout, the standard of construction, sanitation or the environment. As conditions worsened, attempts were made to control them, which led to a series of Public Health Acts. The Metropolitan Buildings Act of 1844 was an important step towards more exacting standards in construction when, for the first time, buildings had to be classified by type rather than size, public buildings had to be fireproofed, and no work could start without prior approval.

The Public Health Act 1875 and the eventual introduction of Model Byelaws in 1877 empowered local health boards to introduce local by-laws to control certain standards for new construction. The Act covered many aspects regarding health and welfare and remained the Principal Act until 1936. Responsibility for the enforcement of the Building Byelaws was entrusted to local authorities on the introduction of the Public Health Act 1936 and eventually, in 1953, Model Building Byelaws with 'deemed to satisfy' clauses which enabled building control to be maintained whilst allowing for new techniques to be introduced by developers derived from R&D. British Standards and Codes of Practice were quoted as suitable standards and have continued ever since. Standards have the advantage of being voluntary agreements by different parties producers, consumers and authorities. Legislation can then be used to define to what extent these standards are mandatory.

The Public Health Act 1961 made provisions for the preparation of the Building Regulations, which were introduced in 1965. These were the first National Regulations controlling building construction in England and Wales. After a number of amendments and metrication they were consolidated into the Building Regulations 1972, but were amended again in 1973, 1974 and 1975. The Building Regulations 1976 came into force on 31 January 1977 and were followed by amendments until new Regulations were introduced in 1985. This process is still continuing with, for example, the use of regulations to control the consumption of energy within buildings and to reduce the production of CO_2 emissions. Other Acts of Parliament and regulations include legislation to ensure the health and safety of construction workers and members of the general public affected by construction operations and enforced by the Health and Safety Executive. Specific acts include the Health and Safety at Work, etc. Act 1974, the Factories Act 1961 and the Offices Shops and Railway Premises Act 1963 (metricated in 1982). From the 1970s onwards, construction has increasingly been influenced by legislation from the European Economic Community.

In addition to the greater emphasis on human and social issues such as Health and Safety and planning, the changes in the regulatory and institutional framework are also characterised by a systems thinking approach. This includes a growing recognition that individual construction projects and the industry as a whole are part of wider environmental systems.

Changes in the technical support infrastructure

Governments are important in that they are clients for public works as well as passing new laws to regulate construction and finance, setting interest rate policies and providing assistance for social housing. They also influence performance and innovation in the industry through the provision of education and the support of R&D activities.

Up until the early years of the twentieth century, the accumulation and sharing of construction knowledge was largely the responsibility of individuals and their professional bodies. With the increasing complexity of construction, the industry had to become more professional and recruit a higher proportion of educated and skilled people in each of the evolving professions. Initially, when young architects joined the industry they trained as articled pupils but from 1894, when the first school of architecture was founded in Liverpool, it became possible to take a full-time course before practising, until, by 1958, a full-time architectural education became almost the only way into the profession. All the engineering institutions were at first reluctant to adopt a formal qualification and examination structure emulating that of the architects. However, this changed in the twentieth century as engineering embraced the explosion in knowledge in science and mathematics during that time. The surveyors also founded organisations to help define and develop their role and emulate the other professions. They founded their professional body, the Institution of Surveyors in 1869. After several changes in title, it became the Royal Institute of Chartered Surveyors in 1946, but it was not until the 1960s that it became a profession for graduates in which a full-time academic education at university or college was the main route to professional status, and it too moved towards a system of examinations and a well-defined route to professional status (Morton, 2002).

During the Fordist era, most of the R&D activities were also funded by the state on behalf of the industry such as the formation of the Building Research Station, which was followed by the Building Research Establishment. Research activities during the early years of industrialisation were initially limited and informal. This became more formalised and systematic with the setting up of the Building Research Station (BRS), with state funding, in 1921. This later became the Building Research Establishment (BRE). Its role was to protect the public interest (safety, hygiene, etc.) whilst undertaking initial research for the industry. Through its recruitment of scientists, technologists and practitioners it also made a major contribution to the development of the intellectual tools of modern construction – codes, standards, performance, concept-based brief and specification, modular and dimensional coordination, building management and building research. The outputs helped the industry to meet the increased demand for construction after 1945 by promoting the features of Fordism, including its specialisation and fragmentation. The number of staff working for the BRE increased steadily from 1945 until it peaked at the end of the 1970s, after which, like most publicly funded institutes, the BRE were forced to cut back on staffing levels following the decline in demand for construction and increased privatisation.

From this outline, it can be seen that the support infrastructure has played a role in supporting innovation in construction. The institutions played an important role in defining and enhancing the roles of individual specialists within an industry organised on Fordist principles. However, they have also sought to strengthen the identities of their own members, claim specific competence and defend their roles in the face of efforts to change construction's operating system. As Bowley (1966) argues, the separation of responsibilities for design from those for production meant that builders could neglect their own education in design and led architects to neglect production. She also went on to argue that the emergence of quantity surveying encouraged designers to neglect the study of estimating and managing costs. The social, cultural and technical separation of the main participants in the industry's supporting infrastructure must be seen at the very least as being unhelpful in attempts to improve construction's operating system during this period.

Conclusion

This overview reveals that the construction industry has evolved and progressed as it responded to changes from the start of the Industrial Revolution to the 1970s. Many of the examples identified demonstrate that pressure for innovation is strongest when there is demand for new types of buildings and structures and new ways of operating.

The growth of industrialisation and urbanisation created new and expanding markets for construction. This increased demand stimulated innovations in the way in which the industry was structured, its operating system and its products. It also led to significant changes in the way the industry was regulated and in the way its support infrastructure evolved to support the main actors. There are examples of areas where the British construction industry was very innovative and exported innovations to other parts of the world. However, there are also many examples where the UK construction industry was slow to exploit new designs, technological inventions and new methods of working.

The organisational system that evolved during this period led to a remarkable and persistent lack of cooperation and integration between the various parties involved in the design and construction of buildings. There were divergent interests between architects, engineers and contractors, which were not conducive to collaborative working, experimentation and innovation. There were few internal mechanisms for innovation in the operating system and most had to be imposed from outside the industry through clients – particularly the government as a major user of its products and services – or developed by its increasingly sophisticated supply systems. Innovation was also impeded by the volatility of overall market demand and changes in demand between different sectors of the market.

The review of contracting strategies and procurement methods shows that, although there was considerable restructuring of the design and construction process, some redefinition of the roles of the main participants, some redistribution of risk, and some changes in the 'rules of engagement', a substantial number of clients and their advisers still remained wedded to the more traditional procurement methods which emerged at the end of the nineteenth century. Indeed, it can be argued that as the number of procurement systems, functions, specialisms and agencies increased, so also did the scope for confusion of responsibility, the opportunities to transfer risk and the growth of adversarial attitudes. Furthermore, the industry failed to integrate the whole of the operating system, with the specialist and trade subcontractors and supply systems being largely excluded from decision-making in relation to the industry's operating system and its contracts and procurement approaches.

The heavy demands for the products of construction coupled with labour shortages during the post-World War II period in particular resulted in a search for new industrialised methods to reduce the amount of site labour and to increase the productivity of the industry in line with Fordism and the mass production model. During this period, construction did change its products and processes and became to some extent part of the dominant technological-economic paradigm. There is evidence that clusters of interrelated technical, organisational and managerial innovations, based on Fordist principles, coupled with ideas imported from other industries, led to improvements in product performance, productivity and a shortening of construction cycles. However, there was also resistance to fully embracing this approach because of conservatism in relation to change and factors peculiar to construction. The early 1980s saw the increasing rejection of Fordism and the emergence of post-Fordism and new challenges for construction.

References

BABBAGE, C. *On the Economy of Machinery and Manufacture*. Knight, London, 1835.

BALL, M. *Rebuilding Construction: Economic Change in the British Construction Industry*. Routledge, London, 1988.

BANHAM, R. *Theory and Design in the First Machine Age*. Architectural Press, London, 1976.

BAYLEY, L. G. *Building: Teamwork or Conflict*. George Godwin, London, 1973.

Bender, R. *A Crack in the Rear-view Mirror: a view of industrialized building.* Van Nostrand Reinhold, New York, 1973.

Bowley, M. *The British Building Industry: four studies in response and resistance to change.* Cambridge University Press, 1966.

Bruland, K. *The transformation of work in European Industrialisation.* In P. Mathias and J. A. Davis (eds.): *The first Industrial revolutions.* Blackwell, Cambridge, 1989.

Burnes, B. *Managing Change: a strategic approach to organisational dynamics.* 3rd edn. Pearson Educational, Harlow, Essex, 2000.

Burns, T. and Stalker, G. *The Management of Information.* Tavistock. London, 1966.

Chapman, S. D. and Chassagne, S. *European Textile Printers in the Eighteenth Century: A study of Peel and Oberkampf.* Heinemann, London, 1981.

Construction Industry Research and Information Association (CIRIA). *Specialist trade contracting – A review.* Special Publication 138, 1998.

Cox, A. and Townsend, M. *Strategic Procurement in Construction: towards better practice in the management of construction supply chains.* Thomas Telford Publishing, London, 1998.

Davey, N. *Building in Britain.* Evans Brothers, London, 1964.

Emery, F. E. and Trist, E. L. The causal texture of organisational environments. *Human Relations*, February, 1965, 21–32.

Franks, J. *Building Procurement Systems: a client's guide.* 3rd edn, Longman, Harlow, 1998.

Galbraith, J. *Designing Complex Organisations.* Addison-Wesley, Reading, MA, 1973.

Gann, D. *Building Innovation: complex constructs in a changing world.* Thomas Telford, London, 2000.

Gann, D. M. and Salter, A. J. Innovation in project-based, service-enhanced firms: the construction of complex products and systems. *Research Policy*, **29**, Elsevier, London, 2000, pp. 955–972.

Glover, J. W. D. and Rushbrooke, W. G. *Organisation Studies.* Nelson BEC Books, 1983.

Guedes, P. (ed). *The Macmillan Encyclopaedia of Architecture and Technological Change.* Macmillan Reference Books, London, 1979.

Herzberg, F., Mausner, B. and Snyderman, B. B. *The Motivation to Work.* Wiley, 1959.

Hibberd, P. R. Key factors in procurement proceedings of CIBW92 – Procurement systems, 1991.

Kast, E. F. and Rosenzweig, J. E. *Organisation and Management. A System and Contingency Approach.* 4th edn, McGraw-Hill, 1985.

Kriedt, P., Medick, H. and Schlumbohm, J. *Industrialisation before industrialisation.* Cambridge University Press, Cambridge, 1981.

Landes, D. *A Characteristic Approach to Technological Evolution and Competition.* Manchester, University of Manchester, 1969.

Lawrence, P. and Lorsh, J. *Organisation and Environment.* Harvard University Press, 1967.

MacGregor, D. *The Human Side of Enterprise.* McGraw-Hill, 1960.

Maslow, A. H. *Motivation and Personality.* Harper Row, 1954.

MAYO, E. *The Human Problems of an Industrial Civilization*. The Macmillan Company, New York, 1933.
MORTON, R. *Construction UK: introduction to the industry*. Blackwell Publishing, Oxford, 2002.
PELLING, H. *A history of British trade unionism*. Penguin, Harmondsworth, 1976.
PERROW, C. *Organisational Analysis*. Tavistock, London, 1970.
PRICE, R. *Masters, Unions and Men*. Cambridge University Press, Cambridge, 1980.
PUGH, D. S., HICKSON, D. J., HININGS, C. R. and TURNER, C. Dimensions of Organisation Structure. *Administration Science quarterly*, June 1969, vol. 13, pp. 65–105.
SAAD, M. The transfer and management of new technology: The case of two firms in Algeria. (Unpublished PhD thesis, Brighton Polytechnic, 1991).
SEBESTYEN, G. *Construction – Craft to Industry*. E&FN Spon, London, 1998.
STONE, P. *Building Economy*. Pergamon, Oxford, 1976.
STROUD FOSTER, J. *Structure and Fabric Part 1*. BT Batsford Ltd, London, 1973.
TRIST, E. The Concept of Culture as a Psycho-Social Process. Paper presented to the Anthropological Section, British Association for the Advancement of Science, 1950.
WOODWARD, J. *Industrial Organisation*. Oxford University Press, Oxford, 1965.
WOODWARD, J. *Industrial Organisation: Theory and Practice*. Oxford University Press, New York, 1970.

Chapter 3

Construction's response to post-Fordism

This chapter investigates the emergence of post-Fordism and its impact on construction. It aims to review the way in which construction has introduced and managed innovation in its products and processes in order to respond to the challenges of the new paradigm. This review is structured around the contextual model for construction described in Chapter 2 in order to identify the key determinants of innovation in construction.

The emergence of the post-Fordism model of organisation

The idea of post-Fordism has emerged as a way of explaining the changes in economic structures and activity that characterised market economies in the later years of the twentieth century. Most observers who have developed the concept identify a sharp turning point in the nature of economic activity from the early 1970s as a result of the oil price shocks and subsequent crises of competition, inflation and recession.

If the central characteristic of Fordism was the mass production of standardised products, post-Fordism has been characterised by:

- increasingly flexible labour processes and markets
- heightened geographical mobility
- rapid shifts in patterns of consumption
- privatisation, deregulation and a reduction in state intervention
- the greater use of programmable Information, Communication Technology (ICT)
- more integrated and holistic approaches
- more responsiveness to the needs of customers
- a greater recognition of knowledge as a competitive advantage and
- the shift in the nature of competitive advantage from the physical to the intangible and from the seen to the unseen.

There is some disagreement as to the extent to which these post-Fordist developments mark a break with the traditional Fordist model. There are three main schools of thought within the Fordist/post-Fordist debate. One school of thought sees the post-Fordist model as a complete and revolutionary break with the past, involving a transformation of production from a social, economical and geographical perspective. The second approach sees the post-Fordist economy as the natural extension of an exploitive tendency in capitalism as firms seek to maintain profitability and reassert control over labour. A third interpretation involves new forms of flexible production and organisations built on more efficient and adapted Fordist practices, which will be the main focus of the investigation in this book. It is important to recognise the cumulative and evolutionary aspect of this latter model as it involves combinations of former and new elements. Whatever theoretical model applies, the transition from Fordism to post-Fordism is made up of a series of developments. A major characteristic of the new model is the move away from high volume mass production to the introduction of variety, customer focus, flexible production systems and mass customerisation. The economies of scale so necessary under the methods of Fordism are now increasingly being replaced by a new capacity to manufacture a variety of goods in small batches, known as economies of scope. Thus, competitiveness is largely determined by the capacity to customise and respond to the needs of customers. Responding to this kind of diversity requires a significant shift from fragmentation to a more holistic approach including the whole organisation. Increasingly this integration is even extended beyond the boundaries of the firm through linkage between customer and supplier organisations in design, purchasing, and distribution (Lamming, 1993). Indeed, enhancement of effectiveness in dealing with customers through actions such as high variety in production and short lead time requires an overall responsiveness on the part of the firm, and, hence, the use of a technology and management based on integration affecting the whole supply system.

In the past, competition and business were essentially concerned with mass production in which markets were considered as relatively predictable. The key factor in competition was essentially based on price factors; that is to say remaining competitive by delivering products and services at minimum cost. Therefore, performance of such systems was mainly focused on rationalisation and reduction of labour costs in order to improve productivity. However, the situation has changed and firms are dealing with continual fluctuations in which the key factor in competition has shifted from productivity alone to an overall responsiveness to changes in the market including quality, delivery time, design, flexibility, new products and services and price. Thus, survival in the current business era depends on reacting more flexibly and adapting more rapidly to customer requirements. This new approach means coming to terms with the idea of greater product and part diversification, higher quality demands and shorter product development and delivery periods. To meet these new challenges, new organisational approaches known as Post-Fordism have emerged and have led to new principles for production, organisation and management marked by empowerment, less division of tasks and more integration of functions and activities.

As already suggested by Piore and Sabel (1984), new production priorities based on non-price factors need an alternative built on flexibility that the mass production organisation cannot offer. Such a new solution needs a closer working relationship between different departments and functions of a business and the dismantling of boundaries between them. Thus, a major prerequisite for effective management is the adoption of a total system approach. This includes the removal of boundaries and the need for cooperation and communication between the different departments of an organisation implies alterations in skills, structures, procedures and even culture (Lui *et al.*, 1990).

Post-Fordism brings a new 'way of thinking' based on changes throughout the entire business. Success in dealing with such significant and fundamental changes calls for a new approach to learning. Indeed the ability to deal with unpredictable tasks and to respond as quickly as possible to changes from the internal and external environment requires knowledge about the system as a whole, taking into account the various interactions. This form of learning must also include understanding about the strategic and organisational implications of this form of management. Learning which encompasses the intra-firm managerial and organisational capabilities is needed to provide an environment generating effective change.

Post-Fordism has led to a change of situation from the point where workers are subordinated to machines and concerned with simple and repetitive operation activities, to a situation where the worker's main task is to monitor and maintain the new technology. This radical change of labour process in the new organisation of work is increasingly being associated with empowerment, including a greater participation, motivation and involvement of the workforce at all levels.

With Fordism and its specialisation and division of work, decision-making is not located where information is available and this often slows the production process and militates against the commitment of the workforce. The participative decision-making associated with post-Fordism is an effective implementation tactic that increases involvement and commitment, key prerequisites for the implementation of innovation (Leonard-Barton, 1991).

The capacity to adapt to changes in the wider environment is now a vital element for a firm's survival. This new need has led to the development of a new production system combining efficiency and flexibility and hence affecting the whole organisation. Organisational changes are concerned essentially with responses to shifts in work organisation, such as those brought about by removing the separation of conception and execution functions and the consequent reduction of hierarchical levels and procedure control. The removal of boundaries between departments and similar activities leads to the emergence of teamwork. Unlike the traditional form of organisation, the coordination between different groups is often performed via informal mechanisms and procedures. With the set up of teamwork there is also a need for new structural arrangements with fewer levels of hierarchy and better communication flows.

The design of new organisational arrangements must be based on the view that employees' motivation and commitment are the predominant constituents for employees' participation, creativeness and innovativeness at work. This is why organisational attributes such as structure, procedures, leadership, control, regulation, job integration and coordination, which were previously based on an entirely different philosophy, need to be reshaped in order to ensure the involvement of the innovative and creative capacities of people at all levels. The handling of organisational factors can facilitate the establishment of a new culture and new values. This culture and these values can help to bind people together with a sense of belonging and a sense of common purpose and can also generate an environment conducive to innovation.

Group work is also playing an important role in the successful implementation of new organisational concepts and in the development of innovation and creativity. It is now widely acknowledged that through group work the employees' know-how is recognised and used to continuously develop their motivation and creativeness, and improve their performance. The cross-fertilisation of ideas, which is the outcome of group interaction, is more likely to influence the importance and originality of the group innovations. A greater degree of self-regulation offered to group members with regard to their work through projects is considered to generate significant employee motivation and participation. A project is in essence an inter-group activity where a sense of common purpose and effectiveness are required.[1]

Communication can also develop and sustain mechanisms for organisational creativeness and innovativeness. The success of new organisational concepts such as Quality Circles, Total Quality Management, Just in Time, Partnering and Supply Chain Management is convincingly

---

[1] Further details of the growing importance of projects and their management is provided in Chapter 5.

associated with the effectiveness of mechanisms of communication and information sharing.

The main objective of the new organisational arrangements is to ensure the employees' commitment and their willingness to identify their individual goals with the organisation's goals (Saad, 1994). This is considered to be attained through the establishment of an organisational culture derived from the combination of strategic and organisational factors.

Chief of all the elements of the post-Fordist approach is the necessity to change the culture of the organisation. Culture indicates a set of common theories of behaviour or mental programmes that are shared by a group of individuals (Hofstede, 1991; Mead, 1998; Cartwright, 1999). For Schein (1996) culture is about basic assumptions, premises, values, ideology, artifacts and creations. Organisational culture can be defined as the collective programming of the mind which distinguishes members of one organisation from another. Denison (1990, 1996) suggests that to be effective organisational culture must be built on the following:

- involvement and participation of the workforce at all levels
- consistency whereby clear guidelines are well understood by the members of the organisation
- communication
- adaptability to respond to external and internal requirements and
- the existence of a well-defined mission which provides purpose, meaning and clear direction.

As has already been discussed, culture is the major factor in improving performance because it promotes new lines of thinking and actions likely to encourage motivation, creativity and innovation. It is, indeed, a careful re-engineering and even reshaping of the way things are done, aimed at improving what works and giving up what does not. Once these new lines of belief, conduct and actions are identified, they need to be learned by the organisation's members (individual learning) and the organisation (organisational learning). Learning facilitates a continuous adaptation to changing circumstances. It is about ensuring that people and their organisations possess the appropriate skills, competence and motivation to be creative and innovative within the context of partnering and other organisational changes within construction projects.[2]

The approaches used by construction up to the early 1980s had not been successful in all the market demands for larger and different types of buildings and therefore not capable of meeting the requirements for reduced cost, greater variety, flexibility and customer focus. Like many

---

[2] The importance of learning and culture in developing innovation in construction are addressed in Chapter 8.

other sectors of the economy, construction has been introducing innovations to reshape its whole organisation and culture to address the new challenges associated with global markets and the increasing use of ICT.

Innovation associated with key features of post-Fordism is beginning to emerge and influence the organisation of construction as illustrated in Fig. 2.1. Gann and Salter (2000), for example, argue that the traditional boundaries between manufacture, design, construction and the service sectors are being eroded. There are also attempts to market demands for a broader and more coordinated range of services including raising of finance, planning and design, consultancy, customer support and training, supply chain coordination, risk management and facilities management. Similarly, there are also changes affecting construction's regulatory and institutional framework and technical support infrastructure.

Construction's response to changes in the construction market

As was seen in Chapter 2, the construction market experienced an unparalleled period of growth from the end of the Second World War to the first oil crisis in 1973. This period of post-war construction was characterised by strong and steady growth in output, which reflected the growth in GDP during this period. The annual average growth of the UK economy between 1950 and 1995 was impressive by its own historical standards. This helped disguise the decline relative to its competitors as, between 1950 and the early 1970s, the UK economy grew at just over half the rate of its major European competitors. This can be partly attributed to the party politics which imposed their own particular burden on British industry, already suffering from the self-inflicted weaknesses of low skills, poor training and inadequate investment (Chandler, 1998).

The late 1960s and early 1970s marked a watershed which saw the disappearance of the main conditions which had underpinned the relative success of the UK economy during the 1950s and early 1960s: favourable labour market conditions and the absence of major shocks. The long period of full employment, rising union membership, and relatively generous unemployment benefits had shifted the balance of power in the labour market, encouraging employees to push for better wages and conditions. The other major factor that undermined the relative stability of the 1950s and 1960s was the oil crisis of 1973/74, when the price of oil trebled. This first oil price shock was followed by a second in 1978/79. These external shocks continued to be felt – to a lesser extent – through the 1980s, which saw a period of high real interest rates and exchange rate volatility.

The 1980s saw a clear improvement in the stability of the UK economy despite a rise in unemployment and although the UK inflation differential with other major economies narrowed after 1983, the inflationary pressures generated by the consumer-led boom in the late 1980s meant that the opportunity to reduce inflation created by the oil price fall of the mid-1980s was missed. The mid-1980s saw the paradox of a flourishing City and cash-rich country accompanied by growing awareness of the basic flaws in the UK's industrial capability. A series of reports identified the UKs deficiencies in vocational education and training, shop-floor skills and management qualifications and training. By the early 1990s, the underlying elements for industrial success – education, training and attitudes – were beginning to be identified. It was only in the wake of the early 1990s recession that a climate of low and stable inflation was re-established (Sentence, 1998).

During the same period, construction output grew at a similar average rate to the wider economy but with more fluctuations (Drewer and Hazelhurst, 1998), as shown in Fig. 1.2 in Chapter 1. The construction market has been particularly volatile since 1973, initially as a result of the uncertainties of the middle 1970s following two major oil crises. The period between 1973 and 1983 was characterised by relative stagnation followed by absolute decline in the recession of the early 1980s following the shift to monetarist policies. The period between 1983 and 1989 saw considerable expansion culminating in the frenetic growth of the late 1980s which was fuelled mainly by the 'big bang' in financial services and the growth in demand for private housing – both stimulated by the Conservative government's policies. This was followed by the recession of the early 1990s as a result of the overheating in the economy in the late 1980s. The late 1990s saw another increase in demand, mainly as a result of Millennium projects.

The first recession brought about by the first oil crisis in 1973 resulted in considerable confusion and uncertainty in all sectors of the economy and the end of the post-war construction boom. Its impact on the UK construction industry was offset to some extent by the emergence of new and expanding global markets for construction skills and products, particularly in the richer major oil exporting countries of the Middle East. The larger general and specialist contractors, design practices and materials producers were able to compensate for the downturn in the UK construction market by exploiting these new opportunities. This marked a significant step in the increasing globalisation of construction as companies from diverse groups of countries cooperated in joint ventures. The growth in construction by the end of the 1980s meant there were more major contractors working in the OPEC countries of the Middle East than in any single industrialised country (Drewer, 1991). The next periods of recession coincided with the second oil crisis and the introduction of monetarism in the late 1970s and early 1980s following the election of the Conservative government in 1979. This manifested itself in a drastic reduction in demand for public sector construction, particularly for new social housing and infrastructure works, and a shift in influence from public sector to private sector clients.

The rapid expansion of financial services (commonly referred to as 'big bang') during the late 1980s stimulated increased demand in the private sector for construction products and services, particularly for prestigious office developments in the City of London. There was also an increase in demand for private housing and large-scale retail developments located on the fringes of towns and cities. This period was characterised by the increasing influence of private sector clients who began to demand high levels of performance in terms of quality and service – particularly in relation to time. As many of these buildings were for prestigious commercial and retail developments they demanded a level of technological sophistication not previously achieved by the UK industry. As a number of these clients operated on a worldwide basis they also began to benchmark the performance of UK construction against that of other countries. This opening up of the UK market meant that international changes began to have a greater direct impact on the products, processes and methods of the UK industry.

In order to meet their increasing demand for more efficient construction, leading domestic clients began to import the emerging international design and management methods including some of the more flexible forms of procurement systems – such as project management and construction management. American designers and construction managers were encouraged to work in Britain by clients dissatisfied with the performance of the UK construction industry. The American approach, which was influenced by the Culture-Excellence perspective, involved detailed design being carried out by specialist contractors and the use of construction management as a separate professional discipline. However, Bennett (1993) argues that the approach was too radical with UK architects being reluctant to relinquish their traditional responsibilities for detailed design and the reluctance and unpreparedness of specialist contractors to assume such increased responsibilities. Consequently, some US companies abandoned the UK market, although some of those who remained were joined by UK firms in designing and constructing buildings to world-class standards, such as Broadgate and Canary Wharf.

The other major influence on the UK construction industry in the post-Fordist era has been Japan. In the early 1980s, Japanese manufacturing companies such as Toyota and Nissan began investing in the UK. These leading companies brought with them their designers and contractors thus exposing the UK construction industry to further international comparisons and competing management approaches. The Japanese approach included their strong focus on long-term relationships, investment in R&D and high quality of product and service. Bennett (1993) argues that the 'added ingredient' provided by the Japanese approach was their emphasis on trust and cooperation rather than confrontation. This approach was adopted by a number of UK clients including Stanhope who were one of the first clients to recognise the benefits of cooperative long-term relationships between customers, designers, main and specialist contractors and suppliers.

The inability of the indigenous UK construction industry to fully meet the surge in demand in the late 1980s – particularly in relation to commercial and residential developments – led to an intensification of the globalisation of the supply side of the industry as materials, components and indeed skills were imported from abroad. The period of frenetic growth during the late 1980s was followed by the severe recession of the early 1990s which, in turn, was followed by a tentative recovery from 1994 to 1996. The last years of the century saw an increasing growth in demand up to and beyond the turn of the century, mainly attributable to the Millennium projects. This demand has continued as a result of the policy of the Labour government to increase public sector expenditure on schools, hospitals and social housing schemes.

As has already been described, the early years of the post-Fordist period saw the growing influence of the regular and frequent private sector client. This growing influence was partly a result of the privatisation of many industries and services during the 1980s by the Conservative government and its deregulation of much of the economy. This greater focus on construction by its major clients was partly attributable to their growing understanding of the contribution that better managed built assets or facilities could play in increasing the competitiveness of their core business. Facilities include buildings and their grounds as well as the utilities and equipment in the buildings such as heating systems, lifts, furniture and computers. This means that facilities are a significant asset of many organisations and can represent, on average, between 32 and 44% of the non-financial assets of an organisation. On the other hand, they can be a major cost to an organisation. In the financial sector, for example, Ballesty (1999) suggests that the costs of premises and offices can account for approximately 20% of total costs. Facilities also have a major impact on the effectiveness of an organisation's people who can account for a further 50% of its costs. Facilities can, therefore, have major direct and indirect impacts on the efficiency and effectiveness of an organisation and hence its competitiveness.

Leading construction clients also began to realise the weaknesses in the design and construction of their facilities. This was due both to their own failings as project owners and sponsors and major weaknesses in their suppliers of construction products and services. More experienced clients, with significant building programmes and, therefore, considerable influence or leverage over their construction suppliers, began to address the poor performance of many of their construction projects, and initiate what they saw as the necessary structural and behavioural changes in their construction supply chains. They increasingly realised the role that a more efficient and effective construction industry could play in helping them respond to the new competitive challenges of post-Fordism in their core businesses. As they became more aware of the nature of the responses needed if organisations were to maintain and indeed increase competitive advantage in the new business environment, so they began, in turn, to demand similar responses from their suppliers of construction products and services. Working with fewer preferred construction suppliers they began to press for more customised products and services, which more closely focused on their requirements. With the development of these longer-term relationships, clients with significant and regular demand for construction products and services began to encourage and support innovation in their supply chains. The needs of clients became more explicit and wide ranging and there was growing emphasis on performance measurement and benchmarking between sectors and countries. In return for more continuity of work and greater certainty of profits they expected their suppliers to deliver innovations aimed at continuously improving project performance through more effective and integrated teams. In other words, they began to demand the same quality of products and services from their construction suppliers that their customers increasingly demanded of them.

During the early 2000s, changes in government spending proposals, particularly in the areas of health and education, began to reverse the decline in demand for public sector construction from the early 1980s onwards, and once again increased the influence of government as a major client as well as a regulator of the industry. This started, to some extent, from 1992 with the increased use of the Private Finance Initiative (PFI), which was introduced by Kenneth Clarke, the Chancellor of the Exchequer under the Conservative government at the time. He argued that 'the private sector will take forward projects which the public used to undertake and do so more efficiently' as private companies could draw on better design and management, enabling projects to be completed more quickly to a better specification, within budget and with less risk to the public sector in the event of project failure.

Although introduced by the Conservative government, PFI was taken up, expanded and considerably refined by the Labour government when it came to power in 1997. After a slow start, the number of PFI projects has increased rapidly in recent years and it is now seen as an important variant of the more general public-private partnerships. There were many teething problems as both the government and the industry learned to deal with the increased complexities, new roles and responsibilities and increased risks associated with the new approach. The criticisms of PFI include the cost and the way it compromises the scope for decision-making by future generations. There is also still conflicting evidence as to its effectiveness in terms of delivering functional and aesthetically pleasing buildings.

What is clear is that this new approach has brought about substantial changes in the way that the public sector procures public services and the whole character of the public sector as a client. It has forced the industry to approach public sector projects in a substantially new way – particularly regarding the way in which the projects are financed and the extent to which they engage with the project and the responsibilities they accept. This has required actual structural changes in the supply side of the industry through the restructuring of firms and the development of consortia and mergers. There has been a shift towards new services, for example, integrating facilities management and maintenance functions. Assembling the expertise required for PFI schemes has also been one of the main drivers of takeovers and mergers of construction organisations on the supply side as they respond to the wider range of expertise and responsibilities required under PFI projects (Bennett, 2000; Morton, 2002).

The modern computer-based technologies associated with the new post-Fordist paradigm also began to impose new demands on building design and construction as, for example, in the case of 'cleanrooms'. The production of advanced products, such as computer chips, demanded that the air in production facilities is kept completely free of pollution that could degrade the quality or performance of the final product. As well as impacting on the design and specification of the facilities used in the production of advanced technologies, the demand for 'cleanrooms' also affected construction processes. The movement of construction personnel, tools and materials in and around such facilities needs to follow strict rules and appropriate cleaning and maintenance procedures. This imposed a discipline and attention to detail not normally associated with construction work and led to the formation of firms specializing in the setting up of such facilities and their operation.

The widespread replacement of small retail outlets with supermarkets operated by few large companies increased the pressure to centralise warehousing and distribution facilities. This, coupled with shortages of land, stimulated the demand for new forms of buildings and handling methods to store goods vertically rather than horizontally. The rapid and relentless growth in travel stimulated construction work in a number of areas. The use of the motorcar increased demand for modern, sophisticated manufacturing facilities, new motorways and roads, motorway service facilities and garages, and the expansion of low-cost flights and tourism have fuelled airport improvements and expansion. In the case of offices, increased space was needed for information technology and conference and meeting rooms. To accommodate the information technology many office buildings were increasingly constructed with raised floors and suspended ceilings for the distribution of cabling, power supplies and air conditioning for the dissipation of heat. Open-plan became more common and as personal space decreased more attention was paid to providing interesting common areas with good air quality. Shell and core is a concept developed during this period. This requires the internal office areas to be left as a shell on completion with tenants finishing them to their own specific requirements. In this way 'offices are no longer seen as factories, but places to share ideas, showcase achievements, do some thinking, humanise technologies and even have fun' (Sir Christopher Frayling).

Although conventional manufacturing activity declined during this period, there has been a steady growth in distribution, warehousing and retail operations. The pressure to centralise warehousing and distribution facilities coupled with shortages of land stimulated the demand for new handling methods to store vertically rather than horizontally. The concept of automating the lifting devices combined with computerised stock control was another logical development and which eliminated the problem of operators working in inhospitable conditions. This resulted in higher and heavier racking systems and storage equipment being used, making greater demands on the design and construction of floors. Once forklift technology had reached its capacity, the stacker crane with vertical masts was introduced and became integrated with the roof and wall structures. Wide spans were accommodated by advances in structural steel design and expansive floor areas through a better understanding of the behaviour of large areas of concrete slab and the introduction of new floor laying equipment and lazer control systems.

The scale of warehouse and production facilities coupled with the speed with which they are required to be constructed increasingly demanded that such buildings be considered as a single interconnected system. In order to maximise efficiency, all the elements – the floor, storage systems, the materials handling equipment and indeed the building structure – are increasingly designed and constructed to common tolerances and performance requirements. A number of main requirements continued to dominate factory design and construction. The provision of flexible space was important as manufacturers sought to optimise production layout and flows of materials. Low-cost and fast construction also remained important. Despite some limited experimentation with tilt-up construction, the current dominant solution comprises single-storey steel frames with lightweight profiled metal cladding and in-situ concrete floors.

From the late 1990s onwards there were once again increasing calls for greater use of prefabrication and off-site production to meet demand in certain construction sectors. There are a number of reasons for this pressure on the industry to reconsider once again the optimum relationship between on-site and off-site production. These included meeting the increasing demand for affordable social housing and more hospital beds. It is also seen as a way of addressing the chronic shortage of skilled construction workers and supporting drives to improve the quality of construction products. As was seen in Chapter 2, the adoption of prefabrication and industrialised or systems building for housing is not new. After the First and Second World Wars the industry came under intense pressure to adopt new methods. In the 1960s, there was considerable enthusiasm for industrialised building systems based on large prefabricated components, which proved to be generally unsuccessful.

Prefabricated timber frames were also used extensively for a time into the 1980s but failed to achieve universal acceptance because of concerns over durability and the quality of on-site workmanship. However, the use of smaller prefabricated components, such as roof trusses, window frames, doors and staircases, continued and expanded in the post-Fordist era. Interest in greater prefabrication increased again in the late 1980s with the introduction of sophisticated components such as external cladding for prestigious office buildings and the introduction of fast-track methods where bathroom pods were introduced to speed up construction. Further interest in prefabrication was stimulated by the report *Rethinking Construction* which, among its recommendations, called for more use to be made of factory-produced components.

During the period of conspicuous consumption during the late 1980s, construction had been, to some extent, distracted from environmental issues. The 1990s saw renewed attention on the impact of construction's products and processes on the natural environment. As was seen in Chapter 2, innovation in building services during the Fordist era was focused mainly on liberating human activities from the restrictions of climate and the seasons. Progress was almost entirely dependent on the massive use of energy consumed in devices to modify and control the environments within buildings. Present-day sophisticated air conditioning is the result of many years of development with the most common systems containing an air-handling unit (chiller, heater, humidifier, dehumidifier), fans, ductwork, coils and filters. The refrigeration used in air conditioning consumes electricity and produces carbon dioxide – the main reasons for its sparing use with the growing realisation of its impact on the natural environment.

As well as concerns relating to the consumption of energy and CO_2 emissions, increasing attention began to be paid to air purity. This was fuelled by the growing complaints and health problems that are caused by air contamination. Increasingly, health is seen as being more than simply the absence of sickness, as it comprises the various components in the relationship between a person and the ambient surroundings. As most people spend more than half their lives indoors, their health is greatly influenced by the characteristics of the enclosed space. These days, it is generally considered more desirable to eliminate potentially harmful materials and gaseous pollutants from within the fabrics of buildings and building spaces as they can have an adverse effect on the occupants, even if present in only very small quantities. Formaldehyde gas is one of the more common emissions from surface materials, and radon is a naturally occurring radioactive gas that can constitute a health hazard at higher concentrations. Recently, increasing attention has been paid to ions in the air, electric and electromagnetic fields and various forms of radiation. As understanding of air quality and its measurement have improved, so performance requirements are replacing ventilation requirements in codes on air quality. In turn, this has stimulated the development of new materials and forms of construction that are less reliant on harmful substances both to the users of the building and construction workers.

With the growing environmental pressures to avoid building on greenfield sites, developers have been forced to build on second-hand sites, many of which had been used for industrial purposes and were therefore often heavily contaminated. This increased use of brownfield sites has given birth to new technologies for cleaning contaminated ground. Concerns over the high level of wastage of materials in the industry and the demand for landfill sites has led to pressure for more recycling of building materials. Consequently, there is now more of an obligation on construction to reduce the volume of construction waste and to avoid dumping it without some useful purpose.

As a result of the shifts in the nature of demand outlined above, construction had to become more adept at meeting the more demanding needs and expectations of individual private sector clients and, more recently, the government. The functions and performance requirements for most types of building continued to change, and indeed accelerate, with the wider adoption of post-Fordism and its new technologies and approaches to organisation.

The growing realisation by increasing numbers of clients of the importance of facilities gave rise to a new function and profession – facilities management – a concept which first appeared in the US and spread to other countries including the UK. Facilities management, which seeks to integrate functions that were previously managed by separate departments or individuals, includes optimising and improving facilities, minimising operating costs and contributing to the effectiveness of the organisation (Haugen, 1994). The emergence of facilities management has resulted in the formation of companies specialising in facilities management and broadened the view of many construction firms.

Specific new designs included modern atria, cleanrooms and integrated electronic systems. Skyscrapers and towers grew even higher and bridges, floors and roofs were constructed with greater spans. New materials and design were introduced for the innovative enclosure of large spaces including domes, space grids and tensioned roofs. Demands for higher levels of finished product performance, the incorporation of ICT and growing environmental concerns stimulated innovations in building design and the materials and components from which they are constructed. The main trends included the use of new materials and materials with new characteristics, new combinations of materials, new structures and components, the reduction of the weight of structures (lightweight construction, curtain walling and light roofs), the introduction of low-maintenance and long-life components, the application of dry assembly methods, further mechanisation of construction processes, and a rekindling of interest in prefabrication and off-site production.

Most of the demand for new types of constructed products grew rapidly from the middle 1980s onwards to accommodate the new economic and social activities associated with post-Fordism and new fledgling technologies of ICT. Advanced and 'intelligent' buildings are terms that originated in the US around 1980 as a consequence of the growth in IT and its increased use in buildings. In an intelligent building, building automation and telecommunication systems are combined into an integrated system to provide the best conditions for the building's occupants and to allow for effective management of resources. The sub-systems of intelligent buildings include building automation systems (indoor climate control, lighting, security and access, fire and maintenance), computer-aided facility management (information systems), internal and external communications (cable networks, TV, fax, electronic mail, telephone, teleconferencing, Intra- and Internet) and internal controls (kitchen functions, business applications).

As well as influencing the design of buildings, IT has also had an impact on the way the development process operates. As drawings providing plans and details of buildings form one of the main ways in which the design of a building is communicated, it is understandable that early computer applications were focused mainly on computer-aided design packages for the modelling of buildings. Programmes have been developed that optimise layouts. Each specialist function within the project team has software to help in their task. This, coupled with the diversity of operating and software systems, has resulted in difficulties in transferring information between different project participants and worked against integration. Modern information management systems are more effective when used as part of a network and when files are saved on a server. More recently, various images and animations can be generated by computer programs. Such solutions are now being termed 'virtual reality' models, which are becoming more and more practical.

The rapid advance of Internet and World Wide Web technologies has resulted in the development of a number of on-line collaboration tools, such as Build-Online, Business Collaborator, and BIW Information Channel. The supermarket chain Sainsbury's has used the Information Channel on over 500 of their new build and refurbishment stores. Most of these systems provide support for information flow, workflow control and process management. There are, however, a number of barriers to the use of these systems. They are difficult to operate in the blame and opportunistic culture which still characterises much of construction. This means much of the information they contain is subject to 'spin'. They use historical and often outdated data. They are still cumbersome and time-consuming, involving much data collection, sorting, maintenance and reporting.

Some office buildings using high-technology components and forms of construction have received international recognition. Notable examples include the Lloyd's headquarters in London, designed by Richard Rogers and Partners with Engineers Ove Arup and Partners. The Hongkong and Shanghai Bank in Hong Kong, designed by Sir Norman Foster, was completed between 1979 and 1985. The latter's precision-made components were sourced from factories around the world and were assembled to very tight tolerances at high speed on a congested site. Among its technological innovations were new mullion solutions for holding the large glass panes, the recovery of heat generated by the computers and a typhoon warning system. Morton (2002) argues that No. 88 Wood Street, in the City of London, is as well designed and constructed as many factory products. The glass cladding is made of special super-transparent glass from France and the 'precision-engineered' lifts were made by Mitsubishi in Japan. Sir Norman Foster used a geodesic dome for the Eden Project in Cornwall which is glazed using triple-layered ethylene tetrafluoroethylene which is one-hundredth the weight of glass.

Despite the growing use of sophisticated prefabricated components, the industry continues to use traditional in-situ materials. This often means that expensive, sophisticated prefabricated components have to be placed alongside traditional bulky in-situ materials. This presents particular challenges for design and on-site construction teams as the forms of construction and their associated processes are very different (Bennett, 2000). This brings two very different production traditions – the on-site craft trades and off-site prefabrication – into close proximity. This presents challenges in forming connections in a way that maintains functionality whilst accommodating very different levels of dimensional accuracy.

During this period, innovations to extend the use and improve the performance of traditional construction materials continued. In the case of concrete, the improvement of concrete-mix design made it possible to achieve much higher compressive strengths and improved reinforcement increased the bond between steel and concrete, reducing cracking and permitting a more efficient use of steel. Additives were developed which, for example, can accelerate or retard setting and improve workability. Accelerating setting and curing times have allowed faster striking times for formwork. This, coupled with the rationalisation of the design of concrete components – particularly in relation to suspended floor slabs – and the development of more sophisticated falsework and formwork systems, has resulted in much faster construction times for concrete structures. Lightweight aggregates can be used to reduce weight and increase thermal insulation properties and aerated concrete has resulted in better insulation properties for concrete blocks and other building elements. In the case of steel, new hot-rolled and cold-formed steel sections became available, and improved steel technologies brought new properties such as higher strength, improved weldability, and non-corroding steels. The susceptibility of steel to corrosion resulted in the increased use of the alloy stainless steel, which is used for a range of building products where long life and zero maintenance are required, such as angles, bolts and tension wires. Other alloys such as Cor-ten, a copper-steel alloy, develops a tenacious oxide coating that obviates the need for painting. As an alternative, many improved paint systems as well as plastic and metal coating systems, including galvanising have been developed to improve the resistance of steel to oxydization and enhance its appearance. The other major weakness of steel – its poor performance in fire – led to the development of more advanced intumescent paints which swell up under intense heat producing a layer of insulation around structural elements. Modelling and quantifying fire and its effects, coupled with a greater knowledge of materials, has led to a new branch of knowledge called fire safety engineering. These developments have allowed designers to use steel framing and other materials in new and innovative ways.

Timber also continues to be used as a result of advances in treatments and its inherent thermal insulation characteristics. The versatility and favourable strength to weight ratio of the material means that it lends itself to prefabrication. It also has strong green credentials because it is a renewable source, which during its growing period helps to reduce CO_2 emissions. Plastic products continue to replace components in traditional materials despite their impact on the natural environment and problems of disposal. Extruded sections are used to produce relatively maintenance-free window frames, doors and frames. Plastic plumbing is more widely used as an attempt to address the shortage of plumbers and long-spanning roofs over large spaces are made increasingly from polymers.

There have been major advances in cladding systems. Much of the surface of many curtain walling systems is glazed, and this has stimulated the development of special types of glass including those that are more effective at controlling internal environments and reducing the financial and environmental costs of heating and air conditioning. Solar control glass is specially manufactured to reduce solar heat gain either by absorption, re-radiation or by reflection, while still allowing a high proportion of visible light to be transmitted. Low-emissivity glass, such as Pilkington's hard-coated K-glass, is usually made with a thin layer of indium tin oxide, with silver or copper-indium tin oxide applied to the inner surface to restrict the passage of radiant heat. Double or triple glazing is increasingly in evidence to reduce energy costs and photosensitive glass, which changes its transparency in response to different amounts of light falling on it, is increasingly used.

As has already been seen, prefabrication of certain building components was part of the fast-track methods developed during the boom in office development in the 1980s. This included the prefabrication of air handling plants which were proving difficult to produce quickly and to the required quality given the difficult working conditions on site. More recent experiments in the prefabrication of other building elements include 30 flats built for the Peabody Trust at Murray Grove in London in 1999. This scheme used prefabricated, room-sized volumetric modules, which were fully finished internally. The modules were produced in a factory in York, delivered to the site by lorry, lifted into position by crane, bolted together and service connections made. The tenants were delighted with their new homes and no defects were reported. However, the costs were 5% above what they might have cost using traditional construction methods.

Despite the difficult site, Raines Dairy in north London is Peabody Trust's follow-up to their acclaimed Murray Grove project and is the UK's largest factory-assembled affordable housing project. This development of 30 flats was constructed using factory-built modules, each complete with a fully fitted interior, that were slotted together on site. Architect Allford Hall Monaghan Morris has designed the six-storey scheme in a T shape. At ground level, eight live-work units act as a buffer against the nearby busy road. Stacked above these are five levels of two-bedroom apartments. To the rear, forming the crossbar of the T, is a wing of three-bedroom family accommodation overlooking a landscaped courtyard.

Adam Preece, Peabody's development manager, puts the case for the approach succinctly.

Murray Grove was a prototype; it was an attempt to see if we could build a housing scheme using modules. For this project, we wanted to take that success to the next level.

Perhaps the biggest change of all, however – and the one that will have the biggest impact on the future success of volumetric construction – is that a different form of construction contract has been used for the Raines Dairy scheme. Murray Grove was procured under a JCT form of contract, which, according to Preece, 'puts a main contractor between Peabody and Yorkon'. For this project, Peabody has changed to the partnering contract PPC 2000. Preece hopes that this will help the Trust learn more about the use of volumetric construction.

At the end of Murray Grove, we found that we hadn't learned as much as we could have done, whereas this contract has got us close to everybody.

For the project's main contractor, Wates, however, the decision to use volumetric construction has helped overcome some of the site's challenges – notably its proximity to a busy railway line and a main road classified as a red route, which means that it has to be kept clear of construction traffic at all times. Using prefabricated modules reduces traffic to and from the site, which in turn minimises congestion and the impact on the site's neighbours. The use of volumetric construction has helped compress the construction programme to 50 weeks after what Preece calls 'a normal lead time' for the project – that is, about 40% less time than it would have taken if the building had been constructed traditionally. It has also allowed construction to take place close to the railway cutting where traditional construction would not have been possible without a series of track closures. But perhaps the biggest benefit in using modules has been for the scheme's future tenants. As Preece points out:

The end users will be getting a superior product, with minimum defects, for less money. It's a win–win situation.

There have been experiments in prefabrication in other sectors of the industry. Early in 2003, the Bristol Royal Infirmary opened two new wards as part of a £32.8 million scheme to boost bed numbers and cut waiting times (*Evening Post*, 12 February 2003). The two new prefabricated wards, designed to accommodate 60 beds, comprise pods made in Teeside and delivered to Bristol, where they were craned into position behind the main hospital building. The steel-framed buildings are expected to last about 60 years and have plastic-coated steel external walls with fibreglass insulation. Patients and staff are reported to be delighted with their new surroundings and that they 'don't feel like a temporary structure.'

Following considerable progress in the use of machines, the industry began to work on the introduction of robots in the late 1980s. Prototype robots have been developed for excavating and moving soil, pouring concrete, bricklaying, slipforming interior finishing and the inspection and repair of wall surfaces. Larger Japanese construction companies have begun developing complete automated construction systems. The Shimizu Manufacturing System by Advanced Robotics Technology (SMART), claimed to be the world's first all-weather automatic system for building. The on-site automatic plant is enclosed by roof and walls and adds floors with the construction proceeding from the ground upwards. Although progress has been made in Japan, the optimism of the late 1980s and early 1990s in relation to robotics and automation has largely evaporated.

With the increased use of contaminated land, new soil cleaning methods have been developed including hydraulic, pneumatic, thermal, chemical-physical or biological methods. Although little construction and demolition materials are being salvaged or recycled at present the volume is growing rapidly and more and more firms are now specialising in reuse and recycling. In addition, the period of life expectancy of building and structures is now more carefully considered and planned in response to environmental concerns. The degradation, maintenance and repair processes are more carefully studied with the objective of achieving optimal economic performance and easier recycling at the end of the useful life of the building.

In terms of innovations in the industry's operating system, the Broadgate project in London, which began in 1985, is often seen as pivotal in the shift from Fordist to post-Fordist approaches. Its significance can be attributed to a number of factors including the influence of American approaches and the combination of management and technological innovations to produce a 'fast-track' approach. The project, which was managed by Bovis-Schal, an Anglo-American consortium adopted fast-track approaches and was completed with around a 30% reduction in time when compared with conventionally constructed office buildings at that time. Essentially, the fast-track approach involved continuing the design of the later elements of the building while construction was started on the earlier elements such as the substructure and superstructure. It also involved the reduction of on-site labour through the greater use of off-site prefabrication. This included structural elements such as precast floors, toilet and bathroom pods, modular plant rooms, lifts and cladding panels, which included some services such as radiators and ducts. Some of the modules were very large and heavy and stimulated innovation in handling equipment and methods such as the use of air skates. Alongside this, technological developments in computer hardware and software were stimulating the greater use of computerized design and management systems to design buildings, elements, sub-elements and components and to schedule and manage construction work on site. Gann and Salter (2000), argue that this demonstrates the shift from an essentially sequential way of working demanded by traditional construction technologies, materials and methods to the concurrent design and assembly of component parts associated with prefabrication and off-site production. This fast-track approach has been cascaded to other forms of buildings including housing.

From the mid-1990s onwards, fundamentally different approaches to procurement began to appear in certain sectors of construction but particularly in the oil and gas industries and parts of the retail sector. Major clients such as BAA and some supermarket chains were showing greater leadership in how their projects were procured and defining new relationships between the main project participants. New ideas from manufacturing began to influence construction including value management, concurrent engineering, just-in-time delivery and reshaping of traditional processes to reflect the move away from an essentially sequential way of working demanded by traditional construction materials and methods to the concurrent design and assembly of component parts. Specific innovations included new approaches to off-site prefabrication, standardisation, the use of new technology, more sustainable construction, respect for people initiatives, partnering, supply chain integration and other areas of process improvement.

The more complex and demanding performance criteria in relation to construction's products and services also included more efficient and more highly serviced buildings. This created the demand for more advanced technological services to be incorporated into buildings and, in responding to these demands, construction had to develop an infrastructure of specialist consultants and contractors focused on their specific needs. This suggests a significant shift towards an increased focus on the customer – one of the key elements of post-Fordism.

The new ways of working associated with post-Fordism were increasingly being adopted by many of construction's clients. As well as improvements in their building products, a number of leading private sector clients demanded improvements in the service provided by construction. They required greater transparency of the development process and this began to confirm many of its weaknesses, including its fragmentation and poor performance, which had been identified in a number of earlier reports (see Table 1.1). As a result, parts of the supply side of the industry have become more customer focused and increasingly innovative, particularly in relation to its inter-organisational relations and operating system. The extent to which innovations have been introduced into the process and its management is discussed in the following section.

Although there has been more diversification in the types of projects, there has been more standardisation of products and processes in specific supply systems. Construction products, market structure and supply systems have also become more specialised. Where previously there had been extensive interchange of services and skills between different types of project, there has been a perceptible move towards further segmentation of the market and increased specialisation. Indeed, in the past, transferring marketing and tendering activities and resources from one sector of the industry to another was often used as a strategy by contractors to maintain their workload during downturns in the overall market and specific sectors as discussed in Chapter 1.

From the above review it can be seen that during this period there have been substantial innovations in relation to materials and components and the forms of construction – particularly in the case of leading-edge projects where the client, architect, engineers and contractors all combined to produce buildings which are unique and highly original. Many of these buildings comprise sophisticated prefabricated components which involve complex assembly operations requiring high levels of skills. Alongside this, there have also been considerable innovations in more generic and standardised types of buildings such as factories, warehouses and large retail outlets.

Operating system

As well as these sharp differences in construction outputs during this period there were also significant changes in the industry's operating system. These included:

- increasing use of ICT
- higher levels of specialisation in the construction process with the higher incidence of specialist contracting
- restructuring of the development process and a redefinition of the roles played by each group of participants based on new organisational structures and management approaches imported particularly from North America and Japan
- globalisation of leading construction supply systems as parts of construction became an increasingly global industry
- significant increases in the levels of productivity.

Like other sectors, construction has benefited from the use of ICT but has also placed a greater emphasis on this type of change to the detri-

Between 1984 and 2000 the Royal Institute of Chartered Surveyors (RICS) has undertaken biannual surveys of the incidence of the use of procurement systems (Davis Langdon Everest, 2000). This biannual survey identifies the changes in procurement methods in building during the 1980s and 1990s. This survey shows that between 1984 and 1993 the use of the traditional system declined from 78 to 50%. Following a slight increase in use of the system between 1993 and 1995, in 1998 it declined further to around 40%. Before the 1980s, few clients procured buildings using management contracting but demand for it increased from a very low base until 1985 when it represented nearly 5% of the value of work undertaken. Demand appears to have declined to 9% in 1987 and recovered to 15% by 1989. Although it declined to 7% in 1995 it recovered again in 1998 to over 10%. The decrease in the demand for management contracting coincided with an increase in demand for the use of construction management. Construction management did not enter the RICS surveys until 1989. From 7% of total value of work in 1989, it increased its market share to over 19% in two years. However, in 1993 and 1994 it represented only around 4% of total value. By 1998, its market share had grown again to around 8%. Design and build grew from a base of 5% of total work in 1984 to 12% by 1987. It continued at this rate until, in 1991, the system increased in popularity and, in 1993, approximately 36% of total value of work was carried out using one of the variations of the design and build systems. This fell back to just over 30% in 1995, but surged to over 41% in 1998. The increase was gained from the traditional system's share of the market, but during the period 1993–95 the traditional system seems to have regained much of the ground it had lost. During the 1980s and 1990s, a significant number of disputes arose from both management contracting and construction management, which led some clients to move away from these procurement systems. Specialist and trade subcontractors were also not keen on this approach, as they believed that management contractors often used their increased power to exploit them downstream in the process.

The same survey shows that, in 1998, there were over 40 different forms of contract in use in construction. The JCT standard forms serve a highly significant proportion of the market. In 1998, 91% of all contracts by number employed a standard form. By value of work the proportion falls to 68%. This variation shows that the higher value construction schemes employ some alternative form of contract such as construction management. The *JCT with Contractor's Design* (1981) was the most widely used form of contract being adopted in 26% of works by value, followed by *JCT Private with Quantities* (1980) which represented over 21% of the market by value. Yet, the NEC contract, which is widely acknowledged as being the most modern and constructive form of contract, accounted for only 0.45% of contracts by value.

There are also a number of standard subcontract forms available for use but it is difficult to identify a regular pattern of contract usage. The most common form of contract is *DOM/1* which is used on around 40% of occasions. In most cases it was with *JCT80* as the main contract, and in most of those cases they involved some form of collateral warranty. However, most of these standard forms do not fully recognise the changing role of the subcontractor, particularly in relation to the growing design input from specialist contractors.

All of the structural changes outlined above are an increasing feature of the construction industry and demonstrate both the industry's desire and capacity for change. The emergence of design and build as a procurement route seeks to address the problems of the traditional separation of design and construction and offers opportunities to bring construction closer to post-Fordism with its growing emphasis on integration. The Private Finance Initiative emerged in response to the poor performance of many public sector projects and government requirements to limit the tax burden yet commission public sector projects in health, transport, prison service and education.

The Interim Report of the Joint Government/Industry Review of Procurement and Contractual Arrangements in the UK Construction Industry (1993), the Latham review, identified a general agreement within the industry that

the route of seeking advice and action from lawyers is embarked upon too readily. While a relatively small number of these legal disputes actually reach formal Court hearings, the culture of conflict seems to be imbedded, and the tendency towards litigiousness is growing.

The report goes on to suggest that

These disputes and conflicts have taken their toll on morale and team spirit.

A survey undertaken by *Building* (1999) shows procurement trends for a sample of clients with an annual construction spend of more than £5 million. The results from data provided from the 34 respondents out of 60 clients eligible to participate show that in 1995, 87% of their projects were procured on a competitive basis, 13% on a collaborative basis and less than 1% on a stakeholder basis. By 1999, the corresponding proportions were 61%, 16% and 23%. The respondents estimate that by 2005 they will still procure 59% of their projects on a competitive basis, 18% on a collaborative basis and 23% on a stakeholder basis.

Table 3.2 summarises the key objectives and features of each of these main procurement routes (Franks, 1998; Rowlinson and McDermott, 1999) associated with post-Fordism and their impact on innovation.

Table 3.2. Review of main procurement methods associated with post-Fordism

Procurement method	Objective	Key characteristics	Impact on innovation
Partnering introduced in late 1980s	Improve performance through greater collaboration and better relationships	Partnering charters and agreements based on longer term commitment (project or strategic) and relationships, mutual objectives, teamwork, disputes resolution, continuous improvement and an open and no blame culture	First attempts have essentially been developed between clients, consultants, main contractors and some key specialist and trade subcontractors Positive role of specialist and trade subcontractors downstream in the process not yet fully exploited
Prime contracting introduced in late 1990s	Deliver best value over the whole life cycle of the project rather than lowest capital cost	Appointment of prime contractor as main interface between essentially public clients and supply chains Whole life cycle costing, supply chain management, value engineering, risk management, long term relationships based on trust, continuous improvement against targets, leadership and learning	Early and on-going involvement of key participants of the supply chain including specialist and trade subcontractors with greater recognition of the contribution they make and their business needs Empowerment of members of the supply chain by prime contractors through clustering and the appointment of cluster leaders

(*continued*)

Table 3.2. (contd.)

Procurement method	Objective	Key characteristics	Impact on innovation
Public Finance Initiative 1992	To deliver public sector projects more efficiently and with less risk To provide improved facilities and services without raising taxes or increasing borrowing	Involves a number of organisations in financing, designing, constructing and running a new hospital, school, prison or other public facility for up to 30 years in return for an annual payment from the client. A key to making this possible is the income stream generated by PFI projects over their lifetime	A major impact on innovation including a requirement that construction organisations take a longer-term view of the client's requirements Life cycle performance and costs of the facility have to be taken into account Service providers are having to become more aware of the issues involved in the maintenance and management of the completed facility All participants have to think and act differently and integrate a wide range of skills, expertise and resources.

As discussed earlier, the main trends in the adoption of procurement approaches between 1984 and 1998 can be derived from biannual surveys undertaken by Davis Langdon and Everest.[3] The traditional system of procurement declined from 78% in 1984 to 57% in 1993, although it rose again slightly to around 60% during the recession of the early 1990s. Demand for management contracting grew to 15% of the market in 1985 before declining to 7% in 1995 but recovering to just over 10% by 1998. The popularity of construction management was short-lived, reaching a peak of 15% before declining to 4% of total

[3] The latest survey available was published in 1998.

value by 1993, although it did recover to just under 8% by 1998. The most significant growth was in design and build which grew from 5% of total value of work in 1984 to 12% by 1987 and over 40% by 1998. This growth in design and build was largely at the expense of the traditional system. In 1998, there had been a marked rise in the use of contractor designed elements of work within traditional contracts with the contractor's designed portion supplement featuring in 48% of contracts using JCT traditional forms. Management procurement routes maintain a consistent proportion of all contracts placed but their regular choice for projects over £20 million increased the proportion of value delivered through these routes.

Although not identified in the Davis Langdon Everest surveys, following the example of the USA, tentative attempts at partnering began to be introduced in the UK from the late 1980s onwards. It was first used most explicitly in the North Sea oil and gas industries driven by a combination of falling prices and high operating costs. It was also adopted by frequent clients in other sectors – notably in retail and the utilities sectors – to develop more stable, long-term and less adversarial relationships with a smaller number of preferred suppliers, provide greater continuity of work, and encourage continuous improvement. Other key objectives include addressing the fragmentation in projects, opportunistic behaviour, power imbalances and lack of equity, high levels of defects, cost overruns, and contractual strategies and disputes (Bennett and Jayes, 1998; Cox and Townsend, 1998).

However, there are both fundamental drawbacks in the concept of partnering (Christopher and Jüttner, 2000), and substantial difficulties in defining (Cox and Townsend, 1998) and implementing it in the context and culture of construction given its transient relationships and over-optimistic or pessimistic views of human nature and motivation (Barlow *et al.*, 1997; Jones and Saad, 1998; Uher, 2000). The fragmented nature of the construction industry with its multitude of organisations and professional groupings provides a challenging context for partnering. A major weakness in the adoption of partnering is explained by the hierarchical nature of construction. Most partnering in construction is currently predominantly initiated upstream by regular and experienced clients (Bresnen and Marshall, 2000), with less attention being paid to the role of specialist and trade subcontractors downstream in the process (Uher, 2000; Barlow *et al.*, 1997). This suggests that the positive role that other participants such as specialist and trade subcontractors can play in projects has not yet been fully explored (Jones and Saad, 1998).

It is only more recently that leading clients have begun to broaden their concept of partnering and to extend it downstream to include their key specialist and trade subcontractors and suppliers through supply chain management. Also, some public sector clients, have been seeking to emulate leading private clients by adopting prime contracting which was piloted in two projects for the Defence Estates, an agency of the UK's Ministry of Defence. It aims to replace short-term, contractually-driven, project-by-project, adversarial relationships with long-term, multiple project relationships based on trust and cooperation. This can be seen as a very ambitious step-change towards a collaborative approach as it includes experiments in re-engineering project and supply chain structures, relationships and processes based on identifying one or more strategic supply partners in each key supply area, and working with them closely to improve the value that they add to the end product, and then delegating to supply partners – the 'first tier' – the management of their own suppliers – the' second tier' (Holti *et al.*, 1999).

The broader view of the operating system demanded by the use of PFI has resulted in a number of innovations. PFI schemes require construction organisations to take a longer-term view as they can be required to run the completed facility for up to 30 years. Bennett (2000) argues that this requires all sectors of the industry to think and act differently. Some contractors anticipate that as much as 30% of their work in the future will be PFI-based. This, coupled with the use of approaches involving some form of negotiation, means that for some main and specialist contractors, contracts won by competitive tender are now only 20% of their workload. PFI projects also require that economic, social and environmental sustainability need to be taken into account and balanced throughout the life cycle of the facility. Organisations also have to be much more aware of the issues involved in the maintenance and management of the completed facility, and this requires the integration of a wide range of skills, expertise and resources. Because contractors are constructing buildings and then running them for 30 years, there should be more of an incentive to make the facility last longer, with less maintenance and using less energy. However, a report prepared for Construction Research and Innovation Strategy Panel (CRISP) at the beginning of 2003 argues that the PFI has so far missed the opportunity to improve sustainability.

Although the above review of contracting strategies and procurement methods shows that there has been some restructuring of the design and construction process, some redefinition of the roles of the main participants including subcontractors, some re-distribution of risk, and some changes in the 'rules of engagement', a substantial number of clients and their advisers still remain wedded to the more traditional, lowest price procurement methods. It can be argued that the role of specialist and trade subcontractors has not been central to the development of construction's contracts and procurement approaches. Furthermore, it has not been covered by the literature relating to their development and use of contracts and procurement approaches.

In addition, there remains a strong reliance on complex and rigid standard forms of contracts as shown in the Davis Langdon and Everest survey (2000). Within the survey sample, 91% of all contracts by number, and 68% by value, employed a Joint Contracts Tribunal (JCT) standard form. These forms, with their provisions for damages or penalties, retention of money and withholding of payment, are seen as being the least conducive to collaboration (Cox and Thompson, 1998). In turn, the standard forms of subcontract used to appoint specialist and trade subcontractors downstream in the process also reflect the terms of the associated main contract. In many cases, the client (to the disadvantage of the main contractor) and the main contractor (to the disadvantage of the specialist and trade subcontractors, often amend these standard forms rendering them even less conducive to downstream collaboration.

The Davis Langdon and Everest survey also shows that the two forms of main contract which are seen as being the most conducive to collaboration, the *New Engineering Contract* (NEC) and *GC/Works/1*, with their provisions for some sharing of rewards and joint working, were only used for 0.45% and 0.1% of projects by value respectively. This suggests that the vast majority of construction's clients still consider it necessary to resort to formal main and subcontracts to control their main contractors and subcontractors downstream in the process. Although some of the industry's variants in procurement systems shown in Table 3.1 have some potential for extending more collaborative approaches downstream in the process, in the majority of cases this has not yet been used to fully involve specialist and trade subcontractors.

As can be seen in Table 3.2, these new approaches to procurement have resulted in some potential for greater collaboration and integration. However, a number of authors remain critical of construction's attempts to reshape its procurement approaches. Cox and Townsend (1998) argue that there has been no theoretical framework underpinning the development of these procurement approaches. This suggests that construction lacks a systematic and strategic approach to change, which can be seen as impeding the cumulative and evolutionary aspect of Supply Chain Management (SCM) relationships as a complex innovation. A further weakness associated with these procurement approaches is that the culture of construction remains essentially adversarial with continuing reliance on price competition and firm contractual arrangements. Most relationships are still largely arms-length and short-term (Cox and Thompson, 1997; Dubois and Gadde, 2000), with a strong tendency towards the use of litigation to resolve disputes (Latham, 1994). Indeed in the case of some new approaches to procurement, such as management contracting, contractual relations have often become even more complex than in the traditional approach and provide further potential for conflict.

Changes in the regulatory and institutional framework

This period saw continued changes in the Building Regulations. The Building Act 1984 consolidated nearly all the previous legislation covering building control and represented the Conservative government's proposals for changes in the regulation of the industry. It included new proposals for optional privatisation, streamlining of the system and the redrafting of the Building Regulations (Polley, 2001). The private option for building control first appeared under the Building (Approved Inspector) Regulations 1985.

Although the Building Regulations 1985 reflected the contents of the 1976 Regulations, their form and arrangement was radically altered. The Schedule 1 requirements were written in general terms requiring reasonable standards of health and safety for persons in or about buildings. They referred to supporting Approved Documents, which, with the exception of B1: Means of escape, were not legally binding. A review of the 1985 Regulations led to more than half the Approved Documents being revised and the passing of the Building Regulations 1991. The Building Regulations (Amendment) Regulations 1994 brought about specific changes including new approved documents for Parts F: Ventilation and L: Conservation of fuel and power.

A number of amendment regulations were issued by the Department of the Environment, Transport and the Regions and the Building Regulations (Amendment) (No. 2) Regulations 1999 resulted in the 2000 edition of Approved Document B. Approved Documents K and N and the approved document to support Regulation 7 were revised and Document M: Access and facilities for disabled people, was applied to dwellings for the first time. The Building (Approved Inspectors, etc.) Regulations 2000 consolidated previous regulations and incorporated minor amendments and additional requirements.

A significant factor in the changes to the regulations was the raising of the target U-values for elements of buildings from their introduction for the first time in 1965. These targets were progressively raised through the 1970s in response to the rapid increase in the cost of oil. Although reductions in energy consumption remain important, the emphasis has now shifted towards the reduction of CO_2 emissions. Proposals to reduce the U-values of windows will have an effect of the design and manufacture of windows and will require double-glazing for all windows and triple-glazing for metal windows by 2004.

As well as the Building Regulations, certain aspects are regulated by other legislation. A growing area for regulation is the impact of organisations on the natural environment. BS 7750/ISO 14001: Specification for Environmental Management Systems now requires that an organisation develops, implements and maintains an EMS to ensure that its activities conform to the environmental policy, strategy, aims and objectives it has set.

The other major area that is so important that it requires extensive regulation is health and safety. Within EU health and safety legislation, Article 118A of the Treaty of Rome 1957, introduced by the Single European Act 1986, requires that member countries 'pay particular attention to encouraging improvements, especially in the working environment, as regards the health and safety of workers'. Directive 89/391/EEC, commonly referred to as a 'Framework Directive', was established and was enacted in the UK by the Management of Health and Safety at Work Regulations 1992 – known as the 'six-pack'. This enacts a set of Directives implementing health and safety law. These deal with minimum health and safety requirements for the workplace, equipment, personal protective equipment, the manual handling of loads and display screen equipment. The sixth Directive within the framework, the 'Construction Sites Directive' is implemented in the UK by the Construction (Design and Management) Regulations 1994 – known as the CDM Regulations. The regulations specifically and clearly emphasise the management of health and safety in the construction processes of planning, design and production. Whereas in the past, the contractor had the major responsibility for health and safety, the CDM Regulations are designed to make all parties contribute to health and safety.

One of the legacies of the Latham report published in 1994 was the 1998 Construction Act. One of the many issues investigated by Latham was the level of disputes between main and subcontractors. The report made a number of recommendations for increasing the fairness in their relationships and for the resolving of disputes. Both of these recommendations were incorporated into the Housing Grants Construction and Regeneration Act 1998, generally known as the Construction Act. Its two main provisions were the introduction of a faster and cheaper system of dispute resolution based on adjudication and the fair payments system, which banned 'pay when paid' clauses. However, disputes are still ending up in the courts and some main contractors have sought to circumvent the banning of pay when 'pay when paid' clauses by substituting 'pay when certified' clauses.

In terms of the institutional framework, the emergence of post-Fordism in the 1980s coincided with a decline in the influence of the professional bodies described in Chapter 2 and the formation of more powerful client and cross-professional bodies such as the National Economic Development Council (NEDO). In 1992, this was superseded by the Construction Round Table, which comprised a small group of major and repeat clients such as BAA, MacDonald's, Whitbread, Unilever and Transco. The publication of the Latham report in 1994 supported the work of the Construction Round Table and encouraged them to push for change in the industry and to publish their 'Agenda for Change'.

The Latham report, which was published in 1994, proved a major catalyst in this process of change and transition from Fordist to post-Fordist thinking. For the first time, a report attempted to estimate a figure for the inefficiency and waste in the industry. It was estimated that across the whole industry the burden of unnecessary cost could amount to as much as £17 billion, and as much as £7 billion in the case of the public sector. Consequently, the report called for a 30% reduction in costs by the year 2000. It made over 30 specific recommendations covering a wide range of issues from tendering to trust funds. In contrast with many previous reports, which invariably ended up gathering dust on bookshelves, the Construction Industry Board (CIB) was set up in 1995 to implement the recommendations of the report. Specifically, the report called for:

- more leadership by Government in improving the client's knowledge and practice in procurement, particularly in relation to briefing the designers and the selection of procurement methods
- a review of the whole design process and addressing the fragmentation between design and construction
- new forms of tendering
- simpler building contracts which should be clearer, more standardised and less likely to lead to disputes
- greater fairness between contractor and subcontractor to remove one of the main causes of disputes
- simpler ways to resolve disputes where they do occur
- a Construction Contracts Bill, to tackle some unfair practices, to introduce adjudication as the normal method of dispute resolution and the establishment of trust funds for payment
- rationalised and improved training and research programmes
- the replacement of joint and several liability by proportionate liability
- compulsory latent defect guarantees for ten years from practical completion for all future new developments.

In 1994, the Construction Clients' Forum was formed with members being drawn from a range of client umbrella bodies, such as the British Property Federation, and major repeat clients such as Defence Estates.

The Government Construction Clients' Panel was formed in 1997 to provide a single collective voice for government procurement agencies and departments. Industry groupings also began to emerge including the Reading Construction Forum in 1995 and the Design Build Foundation in 1997, both of which had strong client leadership.

The report, *Rethinking Construction*, produced by a Task Force chaired by Sir John Egan and published in 1998, reiterated the basic weaknesses of construction, the importance of client leadership in bringing about change and the need for greater integration in the development process. The report stated:

> *We are proposing a radical change in the way we build. We wish to see, within five years, the construction industry deliver its products to its customers in the same way as the best customer-led manufacturing and service industries.*

The report was both forthright in its criticism and radical in its proposed solutions and represented, at that time, the most explicit presentation to the industry of the approaches associated with post-Fordism. It called for a radical reform of the whole procurement process with more emphasis on relationships and less reliance on contracts. It also called for reduced reliance on tendering, cutthroat price competition and more profits for the supply side. There was a strong emphasis on partnering and performance measurement, which had been shown to work in other industries and in construction, as in the case of BAA – Sir John Egan's own company. It is this concept of partnering that has become the main indicator of what has become known as 'Egan compliance'.

The Task Force's ambitions for the industry were influenced by the radical change and improvement seen in other industries. Their report set out an approach whereby substantial improvements in quality and efficiency could be made. It set the industry the challenge to commit itself to change so that, working together, a modern industry could be created, ready to face the future. The report recognised that at its best the industry is excellent with the capability to deliver the most difficult and innovative projects. However, it acknowledged the deep concern that the industry as a whole was underachieving, with too many clients being dissatisfied with its overall performance. It also identified other weaknesses such as the low profitability of many construction organisations and the under-investment in R&D and training.

The report identified five key drivers of change:

- committed leadership
- focus on the customer
- integrated processes and teams
- quality driven agenda and
- commitment to people.

It also drew attention to the need for ambitious targets and effective measurement of performance as being essential for performance improvement. These targets included annual reductions of 10% in cost and construction time and a 20% reduction in defects per year. It advocated a radical change of processes to achieve these targets. These changes should create an integrated project process around the four key elements of:

- product development
- project implementation
- partnering the supply chain and
- production of components.

Sustained improvement should then be delivered through the use of techniques for eliminating waste and increasing value for the customer. It argued that if the industry is to achieve its full potential it needs to change its culture and structure to support improvement. It must:

- provide decent and safe working conditions
- improve management and supervisory skills at all levels
- replace competitive tendering with long-term relationships based on clear measurement of performance and sustained improvements in quality and efficiency.

The report's recommendations have become established as the 5:4:7 mantra of *Rethinking Construction* – 5 drivers, 4 aspects of improving the project process and 7 targets for improvement.

At the heart of the *Rethinking Construction* initiative is the demonstration projects programme from which a number of case studies in this book are drawn. These projects provide the opportunity for leading edge organisations from any part of construction to bring forward schemes that demonstrate innovation and change, which can be measured and evaluated. These are either site-based projects or organisation change projects. At the time of writing, there are more than 400 projects in the programme, with a total value of over £6 billion, 38% of which are from housing and 62% from the rest of the industry. Innovations include off-site prefabrication, standardisation, the use of new technology, sustainability, respect for people activities, partnering and supply chain integration, and other areas of process improvement.

The *Rethinking Construction* report also asked the industry to develop a culture of performance measurement. An industry-wide group, including representatives from both the demand and supply sides of the industry, government and academia, developed a set of simple headline Key Performance Indicators. These are based on the seven *Rethinking Construction* targets but with the addition of other client satisfaction measures. All demonstration projects are required to measure their performance against these KPIs and to report annually on their progress. The DTI also collects data from the industry at large enabling a comparison to be made between the demonstration projects and the industry as a whole.

Directly following the launch of the report, the Movement for Innovation (M4I) and the Construction Best Practice Programme were established by industry together with government to respond to the recommendation for a movement for change. Whilst M4I takes the lead in general construction, the Housing Forum was established to bring together all those within the housebuilding chain in the movement for change and innovation. In March 2000, the Local Government Task Force was set up to encourage and assist local authorities to adopt *Rethinking Construction* principles.

The Construction Clients' Forum published its *Charter Handbook* in which it set out the obligations that define a best practice client in the post-Egan era. It recognises the importance of the leadership role the client needs to play in changing construction. It calls for continuous improvement of culture and relationships throughout the supply chain based on performance measures. The *Handbook* lists the obligations of a Charter client:

- prepare a programme of cultural change with targets for its achievement over a period of at least three years but preferably five years or more
- measure their own performance in achieving their cultural change programme
- monitor the effects of implementing their programmes of cultural change against the national KPIs
- review and amend as necessary annually their cultural change programme in light of what has been achieved.

It requires clients to have procurement processes that deliver the following improvements:

- major reductions in whole-life costs
- substantial improvements in functional efficiency
- a quality environment for end-users
- reduced construction time
- improved predictability on budgets and time
- reduced defects on hand-over and during use
- elimination of inefficiency and waste in the design and construction process.

It argues that clients should always procure through integrated design and construction teams preferably in long-term relationships. Clients should also enforce supply-side reforms in structural relationships, culture, process and outputs by making them a condition of any relationship with the construction industry, and consultants should be an integral part of the industry and the integrated supply chain.

The early years of this century saw the publication of further influential reports. In October 2000, the Prime Minister launched *Better Public Buildings*, commissioned jointly with the Commission for Architecture in the Built Environment (CABE) and the Treasury. In addition, the *Charter Handbook* was published by the Construction Clients Forum and the National Audit Office published *Modernising Construction* in January 2001. In 2002, the Fairclough report argued that investment in R&D is essential to underpin innovation and continuous improvement in construction. In 2002 the report *Accelerating Change* produced by the Strategic Forum for Construction reviewed progress since the publication of the report *Rethinking Construction*.

The *Better Public Buildings* report focused on the functionality of completed buildings and how they enhance the quality of life of the end-user. It also emphasised the importance of achieving better value over the whole life of the building and called for a radical structural and cultural change in procurement practice. It also recognised the important role to be played by specialist suppliers right from the beginning of the design process. It advocated:

- measuring efficiency and waste in construction
- appointing integrated teams focusing on whole-life impact and performance of development
- encouraging longer-term relationships with integrated project teams as part of long-term programmes
- rigorous performance review and
- using whole-life costing in the value-for-money assessment of buildings.

It argued against regarding good design as an optional extra; treating lowest cost as best value; valuing initial capital cost as more important than whole-life cost; imagining that effectiveness and efficiency are divorced from design.

The *Modernising Construction* report outlined how the procurement and delivery of construction projects could be modernised with benefits for all – the construction industry as well as clients. It argued that experience shows that the acceptance of lowest price bids does not provide value for money in either the final cost of construction or the through life and operational costs. It also suggests that relationships between the construction industry and government departments have also been typically characterised by conflict and distrust that have so contributed to poor performance. It called for better integration of all stages in the construction process, much more focus on end-users, less adversarial approaches, the use of longer-term partnering relationships between clients and contractors, improved health and safety, the development of a learning culture in projects and within organisations and the better management of supply chains.

The Fairclough report (2002) was commissioned by and primarily aimed at the government departments responsible for sponsorship and regulation of the construction industry. The main conclusions of the report were presented under the nine headings

- investment
- strategic vision
- mechanisms for change
- commissioning research
- government focus
- responding to unforeseen events
- skills and recruitment
- research base structure, and
- innovative capacity.

On investment in R&D the report recommends that the government should refocus existing resources towards more, better targeted and better utilised work on improving the productivity of the industry and improving clientship, and on strategic longer-term issues. In terms of strategic vision, the review concludes that long-term research planning should be derived from a strategic framework of the issues facing the construction industry. The emphasis should be on key competitiveness and productivity issues and their relationship to achieving sustainability. In terms of mechanism for change, the review concludes that the Strategic Forum should have a pivotal role in setting industry strategic vision and key issues needing action. In terms of commissioning research it concludes that there should be much greater effort on follow through and take up of R&D. The review confirmed the skills and recruitment crisis facing construction. It identified the silo mentality within the industry as a key problem, exacerbated by the stance of professional Institutions and the weak links between industry and academia. It calls for a more interdisciplinary approach to encourage and promote innovation through learning and research.

The report *Accelerating Change,* argued a vision for the UK construction industry realising maximum value for all clients, end-users and stakeholders and exceeding their expectations through the consistent delivery of world-class products and services. It proposed that if the industry is to achieve this it must:

- add value for all its customers
- exploit the economic and social value of good design

- become more profitable and earn the resources it needs to invest in its future
- enhance the built environment in a sustainable way and
- improve the quality of life.

Such an industry, it argues, will be characterised by strong client/customer focused, integrated teams made up of existing integrated supply chains, demonstrating respect for their people (especially in relation to health and safety), working together in a culture of continuous improvement, and with investment in R&D driven by innovation. As well as identifying client leadership, integrated teams and tackling 'people issues' as drivers for change, the report also identifies a number of other cross-cutting issues that can act as enablers or barriers for change. These include sustainability, design quality, IT and the Internet, R&D and innovation and the planning system. Although sustainability did not appear in the *Rethinking Construction* report in 1998, *Accelerating Change* argues that the basic abilities to pre-plan a project through from start to finish is a prerequisite to designing in sustainability and satisfying the triple bottom line of sustainable development, maximising economic and social value, and minimising environmental impacts.

In view of the progress made by *Rethinking Construction* it was given continuing financial support by the DTI for a further two years from April 2002. In order to help to diffuse *Rethinking Construction* across the UK a network of ten Regional Coordinators has been established to manage the Demonstration Project Programme and to work with local organisations to promote the principles of *Rethinking Construction* to the widest possible audience. These Coordinators are working with the industry to develop integrated Rethinking Construction Centres. Their active involvement with the ten Regional Development Agencies (RDAs) and small and medium-sized enterprises, underpinned by the Construction Best Practice Programme, is seen as critical to their success. The four key objectives remain as:

- proving and selling the business case for change
- engaging clients in driving change – demonstrations and Client's Charter
- involving all aspects of the industry – every sector
- creating a self-sustaining framework for changing.

Changes in the technical support infrastructure

One of the main features of this period is the growth in graduate recruitment. However, most undergraduates entering the industry are educated separately and in quite different ways. British architects go through a basic training in schools of architecture in colleges and universities that lasts a minimum of seven years. All courses have to be recognised by the RIBA and are regularly validated. Courses are studio based and teaching involves architectural history and theory as well as the technical subjects such as construction technology, environmental science, structures and materials.

Like the RIBA, the engineering institutions have established a system of qualifying examinations and training based in colleges and universities. After going their separate ways all the various branches now have a common qualification structure. Although academic standards at honours degree level are high, engineers still suffer from a collective inferiority complex and argue that they appear to lack the social standing in comparison with other professions such as medicine and architecture. Surveying is also a well-established profession for graduates from full-time or part-time courses at universities and has managed to raise its status, prestige, its membership and influence in comparison with the other institutions.

Like the other professional institutions, the CIOB has a clearly specified qualification structure and again like them its educational base has to a large extent shifted to the universities and colleges. To grant exemption from the CIOB's qualifying examinations, courses have to be accredited and are regularly monitored. The CIOB's qualifications reflect the structural changes in the industry and are available in five areas:

- facilities management
- commercial management
- project management
- construction management and
- production design management.

The CIOB has been successful in defining a professional basis for construction management and in persuading the industry that its qualifications are a true indication of basic competence. It has over 30 000 members of whom something like one-third are fully qualified. However, although these qualifications are recognised, people can move into management positions through many different routes and from many different backgrounds. In the past it was common for the sons of builders to take over the management of the family firm through what might be seen as a family apprenticeship and this is still common for many smaller firms. Larger firms now tend to recruit graduates from the main construction courses and, more recently, from a wider range of graduates with other professional qualifications including managers, lawyers and accountants.

Universities and colleges have begun to recognise that the problems of fragmentation can start right at the beginning of education and training and some attempts have been made to develop common courses or courses with common modules and early years. The larger multidisciplinary faculties use integrated project work focused on interprofessional issues and working. For example, since 1992 the University of the West of England's Faculty of the Built Environment has had an inter-professional theme through all years of its awards.

As well as concerns over the fragmentation of undergraduate education, there are also more recent concerns in relation to quality and numbers. A 20% fall-off rate in applications since the late 1990s continues a trend started in the recession of the early 1990s. This has prompted universities to accept applicants with lower grades. The Construction Industry Council has warned that the recruitment crisis will affect standards of construction quality, safety, cost and service delivery. This, in turn, will damage the industry's international competitiveness. It recognises that if the industry fails to recruit good quality graduates it will not achieve the improvements set out in *Rethinking Construction*, increase profit margins or improve the knowledge base of the industry. After years of refusing to acknowledge that there was a problem in recruitment, the industry seems to be sitting up and taking notice. Professional bodies, including the Chartered Institution of Building Services Engineers and the Chartered Institute of Building, have established working groups with universities and employers to tackle the issue together. Some industry observers see construction's recruitment crisis as a repercussion of the early 1990s recession, when job insecurity dissuaded students from embarking on long courses in construction disciplines. However, degree courses in certain disciplines including architecture, interior design, town planning and landscape architecture continue to be oversubscribed. This makes another theory more plausible: certain construction disciplines have an image problem among school leavers, and building services engineering, construction management and surveying have been hardest hit.

In terms of postgraduate education, learning on the job and training days still dominate much of construction's approach and there seems to be at all levels a reluctance to accept the importance of formal education and training in improving performance. Indeed, the Egan report identified poor training of supervisors and higher management as being a significant factor in the industry's continuing poor performance. Shared postgraduate training is another route such as the Design Build Foundation's Project Team Leadership Programme, which it has developed in conjunction with Henley Management College. This takes a more holistic, integrated and inclusive view of the development process and recruits students from a range of professional backgrounds including architects, engineers, surveyors and construction managers.

The rise of cross-functional groups to support learning and innovation such as the Construction Best Practice Programme, the Movement for Innovation and the Rethinking Construction movement are all beginning to play a major part in integrating learning by various groups involved in construction. However, not all construction organisations are yet involved – most notably the SMEs that make up the bulk of specialist and trade subcontractors. Also, many of the early events, courses and workshops offered by these groups reflected the lack of understanding of many of the innovations associated with post-Fordism and the fragmented thinking characterising much of construction. More recent offerings such as the Learning by Doing programme of training days offered jointly by the Construction Best Practice Programme and CITB in 2003 recognise the key elements of post-Fordism by including events on Internal Culture, External Relations, and Business Improvement. However, the problems of fragmentation are deep seated and are by no means yet resolved and educating the integrated teams advocated by Latham and Egan still remains a challenge for construction.

Supply systems

Whereas the demand side of the industry is seen as influential in stimulating technological innovation and organisational change, change has also been driven by supply networks. The review of product and process innovations outlined earlier in this chapter have identified the growing significance of the industry's supply networks in supporting construction projects. Most projects use literally hundreds of building materials and, as demonstrated earlier, increasingly sophisticated prefabricated components. In the case of some projects, such as prestigious office buildings, these sophisticated components may be sourced from anywhere in the world. Materials and components will be delivered directly from manufacturers, others will have been supplied through a long and complex supply chain – from production, through factories and maybe builders' merchants.

The unpredictable demand for materials and components and the reluctance of many clients, designers and specifiers to adopt standardised, off-the-shelf components must be seen as a barrier to innovation. It also means that many components have very long lead times as can be seen from the typical lead times for September 2001 and March 2003 shown in Table 3.3.

As well as contributing to the product innovations outlined in earlier sections of this chapter, the supply side has been seeking to apply lean production methods to reduce lead times. Lead times are also dependent on the level of demand. For example, at the beginning of 2003, growing

Table 3.3. Typical lead times (source: Building *14 September 2001 and 28 March 2003)*

	September 2001	March 2003
Cladding – bespoke panellised	42 weeks	41 weeks
Passenger lifts (non standard)	38 weeks	41 weeks
Atrium roofs	32 weeks	30 weeks
Cladding – curtain walling	24 weeks	19 weeks
Electrical package	22 weeks	16 weeks
Generators	21 weeks	21 weeks
Escalators	19 weeks	20 weeks
Metal windows	19 weeks	17 weeks
Precast concrete frames	18 weeks	18 weeks
Structural steel frames	11 weeks	9 weeks
Suspended ceilings	18 weeks	15 weeks
Air-conditioning plant	9 weeks	7 weeks
Piling (rotary)	4 weeks	4 weeks
Brickwork	4 weeks	5 weeks

market uncertainties have resulted in delays to major commercial projects and consequently lead-in times. With the growing use of strategic partnering and PFI work, the issue of lead-in times is becoming less significant as the manufacturers of key components of the building, such as mechanical packages, are involved with schemes from a very early stage. This allows more time to develop the design and value engineer it with other members of the project team often over many months.

Conclusion

Construction's products and processes have become more sophisticated and the pressures on the industry to improve have intensified. There is some evidence to suggest that parts of construction are responding to the new market pressures and the main principles associated with the new paradigm. Although there is more standardisation of product and service, particularly in some leading sectors, there has been an increase in the variety of products and services available.

There has also been some increased concentration of supply as market structures and supply systems are becoming more specialised. Where previously there was movement of resources between different sectors, there has recently been a perceptible move towards more specialisation. These increased variations in resource inputs and outputs in the different construction sectors have reinforced the view that construction cannot be seen as a single industry. Indeed, as construction technologies and skills have become more specialised, the demarcation between the different sectors has increased to the point where the industry should be seen as a set of complementary industries. Clearly, this has implications for innovation and the transferability of 'best practice' from one sector to another.

There is evidence to suggest that this alignment and specialisation of supply systems, coupled with a more integrated and inter-professional approach to education and learning, is beginning to address the industry's long-standing fragmentation problems. The industry's fragmentation is also being addressed by the greater use of more integrated and inclusive procurement approaches such as design and build, partnering and PFI. The adoption of 'one-stop shops' for construction services, currently being developed by Amec and consultants Arup, is another sign of greater integration within the supply side of the industry. Also, the adoption of PFI is contributing to the industry's growing ability to work in a more integrated and inclusive way with a range of professional groups, from financiers to facilities managers. Such procurement approaches, based on an effective flow of information and communication, are providing the means of addressing the increased complexity of the development process in a way that ensures the participants/partners play their appropriate role and do not become subservient to others. Partnering, which is examined in Chapter 6, can improve the effectiveness and cohesiveness of project teams through more stable intra- and inter-organisational relationships.

There is also growing evidence to show that construction has improved its performance since the mid-1990s. There has been an increase in the variety of products and services within certain sectors. Output has risen but employment has fallen. A number of surveys have shown increasing client satisfaction with the performance of the industry. Evidence from demonstration projects is showing significant improvements in performance across a number of performance measures. For example, a survey of regular clients, undertaken by the Construction Industry Board and *Building* in 1999, showed that the performance of contractors had improved by 16% since 1995, the ability to keep to price had improved by 25% and the construction consultancy team as a whole had improved by 10%. Construction consultants' design creativity and ability to innovate had improved by 20%. Clients rated the overall value for money provided by consultants at 18% higher than in 1995. In the mid-1990s, clients seldom gave contractors or consultants marks higher than eight or nine out of 10, but by 1999, one in seven clients gave their contractors 10, and one in fifteen clients rated consultants' performance at 10. Recent demonstration projects under the Movement for Innovation banner are also showing significant improvements in construction's performance.

A further measure of the responsiveness of the industry to change is provided by the effectiveness of the Construction Industry Board (CIB) in implementing the Latham report, *Constructing the team*. The CIB was formed in 1994 to implement the key recommendations made in the report. Although Latham made 30 key recommendations in his report, only one – adjudication pursuant to the Housing Grants Construction Regeneration Act 1996 – has been implemented. This suggests that the CIB largely failed to fulfil its mission of reforming construction legislation, and also raises the question of what happened to the remainder of the Latham report. However, the burden of resolving such disputes has, to a certain extent, been shifted away from the courts to adjudicators. This is arguably the significant success story of the *Constructing the Team* and the CIB's work over the past six years. Although there has been considerable progress in relation to more leadership by government following the election of the Labour government in 1997, and simpler ways to resolve disputes, other Latham recommendations have proved more problematic. For example, the widespread use of simpler building contracts, new forms of tendering, and rationalised and improved training programmes. Also his recommendations in relation to liability law and latent defects have done little to change the working practices and adversarial relationships in the industry – the things that Latham considered were causing poor performance and needed to be changed to improve productivity and competition.

Although Latham advocated the introduction of a compulsory latent defect system comparable to that operated in France, clients have not embraced the concept of compulsory latent defect insurance. This is partly because the insurance market does not offer comprehensive policies – particularly in relation to M&E works – and partly on the grounds of cost. The British Property Federation and the Association of British Insurers have been working together to promote a broader range of latent defect insurance policies for clients, but both are opposed to the introduction of a compulsory regime as recommended by Latham. However, unless such insurance is made compulsory, clients are likely to continue to rely on contractors' balance sheets, as they remain a more attractive (and cheaper) alternative.

The CIB has now been replaced by Egan's Strategic Forum and is not addressing many of the issues identified by the Latham report. As a result, a number of the problems in the industry persist and the potential for disputes in relation to construction contracts – one of the areas Latham was most keen to see improved – is as strong as ever. It is still too early to fully evaluate the effectiveness of the bodies set up to drive forward the changes advocated in the Egan report, *Rethinking Construction*.

Despite this considerable pressure for change from the regulatory and institutional framework, much of the industry is still continuing to be structured and constrained by the economic, social and technological trade-offs defined by construction's contracting system that emerged over 150 years ago in the early part of the nineteenth century. Construction remains labour-intensive with major difficulties in adopting more off-site production and productivity growth has been slower than in many other sectors of the economy, where innovations such as SCM have resulted in more substantial improvements in performance and effectiveness. This would indicate that much of the industry remains untouched by post-Fordism as defined in the introduction of this chapter. Chapter 4 reviews the theory of innovation in order to gain a deeper understanding of the innovation process and its key determinants.

References

BALLESTY, S. Facility quality and performance. In BEST, R. and DE VALENCE, G., (eds). *Building in Value.* Arnold Publishers, London, 1999.

BARLOW, J. *et al. Towards Positive Partnering: Revealing the Realities in the Construction Industry.* Policy Press, Bristol, 1997.

BENNETT, J. Anniversary issue of *Building.* February 1993, pp. 183–184.

BENNETT, J. and JAYES, S. *The Seven Pillars of Partnering: a Guide to Second Generation Partnering.* Reading Construction Forum, 1998.

BENNETT, J. *Construction – The Third Way.* Butterworth-Heinemann, Oxford, 2000.

BRESNEN, M. and MARSHALL, N. Partnering in construction: a critical review of issues, prelims and dilemmas. *Construction Management and Economics,* **18**, 2000, 229–237.

BURNS, T. and STALKER, G. *The Management of Information.* Tavistock, London, 1966.

CARTWRIGHT, J. Cultural Transformation – Nine Factors for Improving the Soul of your Business. *Financial Times.* Prentice Hall, 1999.

CHANDLER, SIR JOHN. The political framework: the political roller coaster. BUXTON, T., CHAPMAN, P. and TEMPLE, P. (eds). *Britain's Economic Performance*, 2nd edn. Routledge, London, 1998.

CHRISTOPHER, M. and JÜTTNER, U. Developing strategic partnerships in the supply chain: a practioner perspective. *European Journal of Purchasing and Supply Management*, **6**, 2000, 117–127.

COX, A. and THOMPSON, I. 'Fit for purpose' contractual relations; determining a theoretical framework for construction projects. *European Journal of Purchasing and Supply Management*, **3**, 1997, 127–135.

COX, A. and TOWNSEND, M. Strategic Procurement in Construction: Towards Better Practice in the Management of Construction Supply Chains. Thomas Telford, London, 1998.

Davis Langdon and Everest. *Contracts in Use: a Survey of Building Contracts in Use During 1998*. Davis Langdon and Everest, London, 2000.

Denison, D. *Corporate Culture and Organisational Effectiveness*. Wiley, 1990.

Denison, D. What is the difference between organisational culture and organisational climate? A native's point of view on a decade of paradigm wars. *Academy of Management Review*, **21**(3), 1996, 619–54.

Drewer, S. National and International Approaches to Construction. *Proceedings of the Conference, Quality, Productivity and Environment in Construction*. 25 June 1991, MACE Centre, Imperial College, London.

Drewer, S. and Hazelhurst, G. *Construction Towards the Millenium: A Case Study*. Faculty of the Built Environment, University of the West of England, 1998.

Dubois, A. and Gadde, L. Supply strategy and network effects – purchasing behaviour in the construction industry. Supply Chain Management in Construction – Special Issue. *European Journal of Purchasing & Supply Management*, **6**, 2000, 207–215.

Emery, F. E. and Trist, E. L. The causal texture of organisational environments. *Human Relations*, February, 1965, pp. 21–32.

Ettie, J., *Taking Charge of Manufacturing*. Jossey-Bass, San Francisco, 1988.

Fairclough, Sir John. *Rethinking Construction, Innovation and Research: A Review of Government R&D Policies and Practice*, 2000. DTI/DETR.

Franks, J. *Building Procurement Systems*, 3rd edn. Longman, Harlow, Essex, 1998.

Galbraith, J. *Designing Complex Organisations*. Addison-Wesley, Reading, MA, 1973.

Gann, D. M. *Building Innovation: Complex Constructs in a Changing World*. Thomas Telford, London, 2000.

Gann, D. M. and Salter, A. J. Innovation in project-based, service-enhanced firms: the construction of complex products and systems. *Research Policy*. Elsevier, London, 2000, 955–972.

Glover, J. W. D and Rushbrooke. W. G. *Organisation Studies*. Nelson BEC Books, 1983.

Groak, S. *The Idea of Building*. E & FN Spon, London, 1992.

Haugen, T. Total build – an integrated approach to facilities management, FM in Scandinavia. In: *Architectural Management: Practice and Research*. Proceedings of CIB W 96 Symposium in Nottingham, UK, 17–18 September, 1994.

HM Controller and Auditor General. *Modernising Construction*. HMSO, London, 2001.

Hofstede, G. *Culture and Organisation – Software of the Mind*, McGraw Hill, 1991.

Holti, R., Nicolini, D. and Smalley, M. *Prime Contracting Handbook of Supply Chain Management Sections 1 and 2*. Tavistock Institute, London, 1999.

Jones, M. and Saad, M. *Unlocking Specialist Potential: A more participative role for specialist contractors*. London. Thomas Telford Publishing, 1998.

Lamming, R. *Beyond Partnership: Strategies for Innovation and Lean Supply*. Prentice Hall, New York, 1993.

LANDES D. *A Characteristic Approach to Technological Evolution and Competition.* Manchester, University of Manchester, Mimeo, 1969.

LATHAM, Sir M. *Constructing the team, final report of the Government/Industry review of procurement and contractual arrangements in the UK construction industry.* London. HMSO, 1994.

LAWRENCE, P. and LORSH J. *Organisation and Environment.* Harvard University Press, 1967.

LEONARD-BARTON, D. The role of process innovation and adaption in attaining strategic technological capability. *International Journal of Technology Management, Special Issue on Manufacturing Strategy.* 6:3/4, 199.

LIU, M. Organisation design technological change *Human Relations*, **43**, No. 1, 1990, pp. 7–22.

LOCKE, E. A. *Toward a Theory of Task Motivation and Incentives, Organisational Behaviour and Human Performance.* May 1968, pp. 157–189.

MAYO, E. *The Human Problems of an Industrial Civilization.* Macmillan, New York, 1983.

MEAD, R. *International Management*, 2nd edn., Blackwell Business Publishing, 1998.

MORTON, R. *Construction UK: Introduction to the Industry.* Blackwell Science, Oxford, 2002.

PERROW, C. A Framework for the Comparative Analysis of Organisations. *American Sociological Review*, **32**, 1967.

PIORE, M. and SABEL, C. *The Second Industrial Divide.* Basic Books, New York, 1984.

POLLEY, S. *Understanding the Building Regulations*, 2nd edn. E & FN Spon, London, 2001.

PUGH, D. S, HICKEN, D. J. and HININGS, C. R. *Writers on Organisation – Structure of Functioning of Decision-Making in Management of People in Society.* Penguin Books, 3rd edn., 1983.

ROWLINSON, S. and McDermott, P. (eds). *Procurement Systems: A Guide to Best Practice in Construction.* E & FN Spon, London, 1999.

SAAD, M. and JONES, M. New organisational and cultural arrangements in the management of projects in construction in *Vision To Reality.* Australian Institute of Project Management, National Conference, 1995.

SCHEIN, E. H. Three cultures of management: the key to organisational learning. *Sloan Management Review*, **38**(1), 1996, 9–20.

SEBESTYEN, G. Construction – Craft to Industry. E&FN Spon, London, 1998.

SENTENCE, A. UK macroeconomic policy and economic performance. BUXTON, T., *et al.* (eds). *Britain's Economic Performance.* 2nd edn. Routledge, London, 1998.

UHER, T. E. Partnering performance in Australia. *Journal of Construction Procurement*, **5**, No. 2, pp. 163–176, 2000.

WOODWARD, J. *Industrial Organisation.* Oxford University Press, Oxford, 1965.

Chapter 4

The conceptual framework for innovation

This chapter explores and analyses the ways innovation has been addressed and defined. It examines the major attempts to explain innovation as an evolutionary, dynamic, lengthy and complex process. A greater emphasis is placed on identifying the innovation process, its main determinants (enablers and inhibitors) and the main models for innovation. This investigation provides a conceptual framework for the implementation and management of innovation in construction.

Nature of innovation

This book views innovation as a new idea which can lead to enhanced performance. For Rickards (1985), innovation is any thought, behaviour or thing that is new. Innovation is not a single nor an instantaneous act but a whole sequence of events that occurs over time and which involves all the activities of bringing a new product or process to the market. The scope of interaction and complexity of innovation is better reflected by the work of Rothwell (1992) and others for whom innovation is an interplay of products, processes and related human behaviour. Rickards argues that innovation consists of two sub-systems. The first relates to the firm and its capacity to deal with innovation and the second comprises the technological, economic, social and institutional factors which form the external environment. It is regarded as a complex sequence of events involving many different functions, actors and variables and forming a process which is not reducible to simple factors. Tidd *et al.* (1997, 2001) claim that failing to secure an understanding of the process of innovation as a whole can lead to one-factor analysis based solely on one of the many important aspects of the process.

Innovation is relevant to disciplines as diverse as engineering, anthropology, sociology, psychology, organisation theory, economics and political science. Each discipline scrutinizes the innovation process according to its own interests. For economists, innovation is viewed as a vital factor in fostering economic growth. For psychologists and specialists of organisational behaviour, innovation is mostly perceived as a team effort and the aim is to understand changes in human behaviour. Sociologists and organisation theorists also focus on human behaviour but only on that part which involves interaction among people in a group or organisation. Their objective is to investigate the role of organisational structure and its appropriateness to promote innovation. Burns and Stalker (1966), Woodward (1965), and Trist (1981) focused on patterns of organisational adaptations and showed that the best structure for an organisation dealing with innovation is one which matches the environment. Similarly, Lawrence and Lorsch (1967) claimed that in a fast changing environment, innovation is fostered through effective communication based on highly differentiated structures and integrated mechanisms.

Political sciences focus on the formation of groups and the way they pursue their interests. They investigate issues including the impact of government decisions, decision-making methods and the politics of implementation. Decision processes play an important role in innovation as organisations and groups are faced with difficult choices to innovate or not, to select from different innovations and methods of implementation, and the associated uncertainty and risk. For engineers and technologists, the primary concern is the design of product and its production process. They focus on the efficiency of the various means for achieving production. Individuals and groups are seen as components to be integrated with mechanical parts into that process. The involvement of each of these disciplines described above is necessary but none of them can, by itself, help gain a thorough understanding of such a complex and multidimensional process. The need for a holistic approach to understand the complex nature of innovation is also demonstrated through a review of the historical development of our understanding of innovation.

Historical evolution of innovation

Classical economists, from Adam Smith to Milton Friedman, have presented the history of innovation as a continuous process of advance. In general, economists have always recognised the importance of innovation in the promotion of economic growth. For Adam Smith in 1776, the division of labour and improvements in machinery led to the promotion of invention. Marx, in 1848, stated the necessity for the bourgeoisie to revolutionise constantly the means of production. It was, however, Schumpeter and Schmookler who identified innovation as an important component of economic development.

In a paper published in 1952, Schmookler stated the belief that the growth in the national product of the USA in the seventy year period leading up to 1938 was due not only to the growth in the stock of capital and labour, but also to the growth of efficiency in the use of these resources. Kuznets (1965) supported this view, which found its greatest advocates in Solow (1957) and Kendrick (1984). They also claimed that from 1900 to 1920 productivity improvement contributed 1% a year to the rise in the national output of the USA, and between 1920 and 1950 this contribution increased to 2% per year. Solow highlighted that no more than 1/8 of the growth of output per head could be attributed to increased capital input, and the remaining 7/8 should be credited to innovation. Thus, the major breakthrough was the identification of the link between growth in output and innovation.

The history of innovation appears as one of steady progress from the time of the Industrial Revolution. As described in Chapter 2, the primary Industrial Revolution, concerned with the development of steam energy, was followed by a series of secondary revolutions (railway, electricity and electronics). One of the first and major contributions to this theory of long-term economic development and structural change was identified by Joseph Schumpeter.

Schumpeter's theory of innovation

Schumpeter's theory of economic development introduced the idea of innovation as an evolutionary process which represented a radical departure from neo-classical economics. The primary concern of the neo-classical school was to explain and predict changes in the relationships between static economic variables such as prices, output and profit in the general equilibrium model based on the assumption of maximisation and perfect competition. For Schumpeter, innovation was the main driving force of growth and decline of economies. In his view, the way economic systems responded to changes was evolutionary, in which innovation and the effectiveness of the entrepreneur as an agent of change played the most significant roles. He also added that economic systems such as capitalism can never be stationary. Thus, Schumpeter was a strong advocate of the theory that the supply of innovation is more important than adaptation to existing patterns of demand. He viewed the entrepreneur as being the key agent in capitalist societies, often discovering new ideas and introducing them into economic life. However, after having identified the individual entrepreneur as the source of innovation in his early work, Schumpeter argued in his later works (1928 and 1939) that it is the entrepreneurial function which generated innovation rather than the individual entrepreneur.

In his study of business cycles (1939), Schumpeter recognised that every stage of the innovation process depends on preceding development and every stage creates the prerequisites for the following stages. These notions of cumulative and prerequisite development demonstrate Schumpeter's acknowledgement of cycles in economic development. The combination of Schumpeter and Kondratiev's work led to the idea of innovation as the basis of the 'Kondratiev cycles'.

In the 1920s, Kondratiev was among the first to study and explain the turning point of the long cycles. For Kondratiev, these cycles were essentially related to the durability of certain types of investment such as buildings and transport. However, Schumpeter identified innovations as causing economic fluctuations. He also claimed that these innovations are not evenly distributed through time but emerge discontinuously in groups or 'swarms'. Schumpeter identified the following waves:

(*a*) the first 'Kondratiev wave' from 1785 to 1845 and which corresponds to steampower
(*b*) the railways as the second 'Kondratiev wave' from 1845 to 1900
(*c*) the third 'Kondratiev wave' from 1900 to 1950 corresponding to electric power and the automobile.

With the emergence of mass production and IT, the fourth and fifth waves were added to Schumpeters' three waves, as shown in Figs 4.1, 4.2 and 4.3.

In addition, Schumpeter was also influenced by the Walrasien theory of 'equilibrium' and started his analysis from a state of equilibrium which gets thrown into disequilibrium. His view of innovation generating disequilibrium led to economic cycles being divided into four phases: prosperity, recession, depression and revival. The prosperity phase is characterised by successful innovations where uncertainty and risk are limited. As a consequence, a shift from equilibrium occurs. When this innovative impetus ends, there is a shift to the recession phase with a new equilibrium. The depression phase appears when the previous innovation has lost its effect. Finally, the revival phase emerges when new innovations are introduced into the market. For Schumpeter, the disequilibrium caused by innovation opens up new opportunities for adaptation which leads to new innovation.

While Schumpeter's contribution to the understanding of the role of innovation in the process of economic change is widely acknowledged, it can be argued that he ignored the process of innovation itself. Usher (1954) and Strassman (1959) argue that Schumpeter's work was merely a description of the consequences of innovation and did not explain the process. Ruttan (1959) and Freeman (1996) assert that he had very little to say about the origin of innovations and management of innovation.

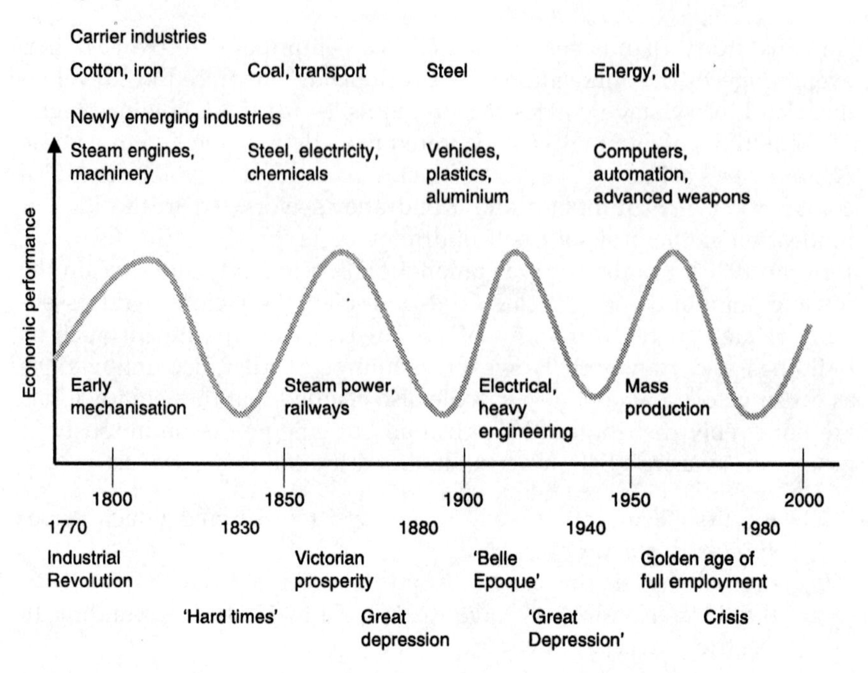

Fig. 4.1. The long cycles of economic performance – carrier industries and newly emerging industries (derived from Bessant, 1991)

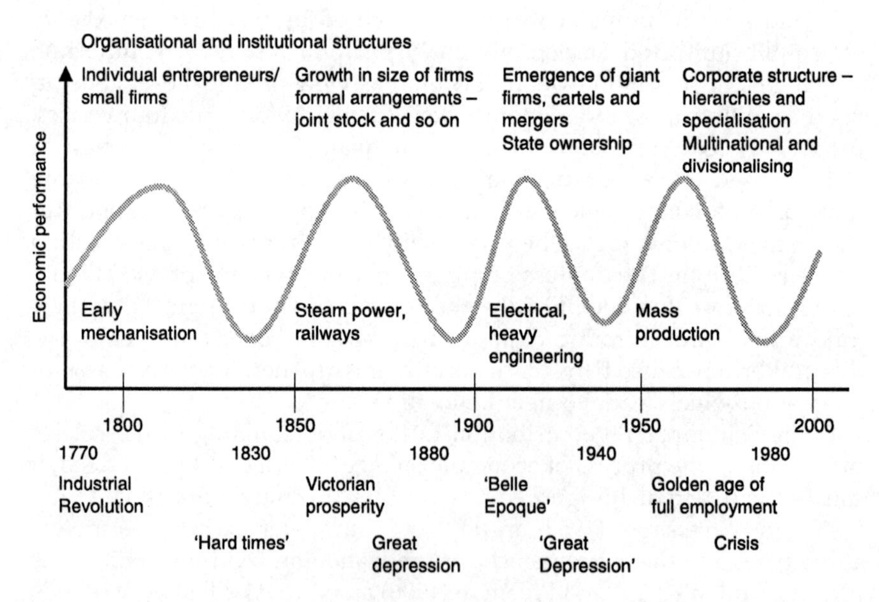

Fig. 4.2. The long cycles of economic performance – organisational and institutional structures (derived from Bessant, 1991)

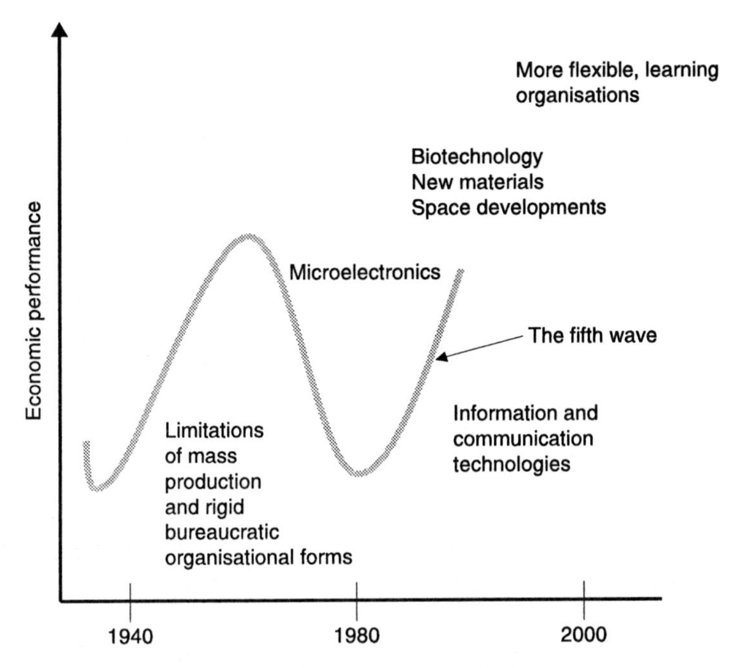

Fig. 4.3. The fifth wave (derived from Bessant, 1991)

Major neo-Schumpeterian contributions to the theory of innovation

The Neo-Schumpeterian work can be divided into three main groups. First, a group where the emphasis is placed on long economic cycles. The second group focuses on search and selection mechanisms and the third group advocates a more organic approach to innovation. For all three groups, innovation, which is the major factor generating economic change, is an evolutionary process which is interactive, cumulative, institutional and disequilibrating.

Long cycles theory

Mensch (1979) proposed a theory of 'bunching' of basic innovation to explain how Kondratiev waves start. He claims that, historically, such a pattern can be observed to occur in the depression periods of the 1830s, 1880s, 1930s and he predicted a new cluster of basic innovation in the 1980s. The cycles are associated with a cluster of basic innovations which stimulate the development of new branches of industry. The resulting economic expansion reaches a limit and consequently an innovation stalemate is reached. This stalemate creates an 'accelerator mechanism' and hence induces innovations which come again in clusters and boost the economy.

This theory was challenged by Freeman and his colleagues (1982) who argue that once swarming starts it has significant multiplier effects in stimulating further demand in the economy. This induces a further wave of process and applications innovations, and gives rise to demand for capital goods, materials, components, distribution facilities and labour. It is this effect of swarming that leads to economic expansion, and not depression inducement. However, Freeman and his colleagues acknowledge that depression may bring about institutional, political and social changes which can facilitate the adoption of new systems of innovation.

Freeman and Perez (1988) introduced the concept of a techno-economic paradigm which is much wider than clusters of innovations or even innovation systems. This concept is based on Kuhn's ideas on paradigms (1962), who introduced new ways of looking at the development of knowledge in general, and scientific knowledge in particular. He claimed that scientific knowledge did not develop in a linear progression but in a series of stages of stability and disruption leading to periodic paradigm shifts in which the whole structure of how scientists see the world is altered. However, Kuhn did not examine the socio-economic context which shapes the emergence of a particular paradigm.

For Freeman and Perez (1988), the concept of a techno-economic paradigm refers to a combination of interrelated products which include step changes in productivity for many sectors of the economy and the creation of an unusually wide range of investment and profit opportunities. Thus, each type of economic development would be influenced by a specific technological style or paradigm (Freeman, 1996). By technological style they mean an ideal type of productive organisation or a best practice pattern which develops as a response to external pressures. This assumes a strong feedback interaction between the economic, social and institutional spheres which generates a dynamic complementarity centred around a paradigm. This is why Freeman and Perez (1988) suggest that the favourable conditions for their concept of paradigm should include adequate complementarities between innovation and the external environment which includes appropriate infrastructure, political stability and institutions that do not substantially hinder the diffusion of new innovation. This provides a better understanding of the main determinants of innovation from within both the internal and external environment, as shown in Fig. 2.1 in the context of construction. Such a paradigm does not only lead to the appearance of a new range of products, services, systems and industries, it has a direct or indirect influence on the behaviour of a significant part, if not the whole, of the economy.

This techno-economic paradigm consists of a combination of radical and incremental innovations and its emergence is basically justified by the scope of changes brought about by these types of innovation. Radical innovations are associated with Research and Development (R&D) activity and are discontinuous events. Incremental innovations occur more or less continuously, depending on a combination of demand pressures, socio-cultural factors and technological opportunities. The incremental innovations occur essentially as the outcome of simple improvements suggested by users.

Freeman and Perez's (1988) techno-economic paradigm is also based on organisational and managerial innovations since the changes involved affect not only engineering activities but also various facets of management activities such as production, distribution and organisation. Thus, each wave is not only concerned with technological clusters but has also dominant organisational forms associated with it. For instance, the fourth wave is associated with organisation based on division of labour and rigid bureaucratic style. The techno-economic paradigm advocates the need for a greater compatibility between the emergence of new technology or innovation and the organisation in which it is to be operated. This explains Bessant's (1991) suggestion of a fifth wave in which emphasis is put on the following features:

- non-price factors
- flexibility in technology
- flexibility in organisational structure and
- changing relationships within and between organisations.

Search and selection theories

The works of Nelson and Winter (1977, 1978, 1982), which contributed to a better understanding of innovation at the organisational level, are acknowledged as being of notable value in the search and selection approach. They define innovation as a change of decision rules within organisations which are more likely to be stimulated by threats and adversities rather than a situation characterised by favourable outcomes. Organisations that are sufficiently profitable do not search for alternative techniques. These 'decision rules' are known as technological trajectories, that organisations may embrace and change depending on the characteristics of their products, processes and the environment in which the organisation operates. The notion of search and selection assumes the pre-existence of possibilities to innovate and rests on the three following concepts.

- Organisations have a set of organisational routines which set out what is to be done and how it is done. This routine is considered to be the genetic code of the firm, it stores information and is used for search processes. These codes carry the adaptive information required for competition and survival. This information in the genetic code changes over time with experiences.

- Organisations undertake a search process for possible modifications or replacements which generates innovation.
- The third concept is the selection environment, which includes all the factors which influence the well-being of the organisation and covers both the conditions prevailing outside the organisation including the behaviour of other organisations.

Nelson and Winter contend that the choice of innovation is basically purposive but they consider the generation of innovation as rather stochastic or random. However, Nelson and Winter's paradigm is defined not only in terms of opportunity and the appropriateness of the conditions, but also in terms of cumulativeness of innovation and in terms of learning procedures. For Pavitt (1984a, 1985, 1987, 1988), Nelson and Winter (1982) and Rosenberg (1979 and 1982), organisations are more likely to search into areas which are proximate to their accumulated experience and capabilities. Lundvall (1990) and Nelson (1993) describe this selection environment as a 'national system of innovation'. This describes the complex nature of institutions and policies which influence and shape the process of innovation and technology at micro-level in any particular national economy.

The innovation approach adopted by a firm will be dynamic and shaped by new internal and external requirements and will therefore not always be similar to those adopted by other firms operating in the same industry or environment. Each situation can generate its own approach, taking into account interactions between internal capabilities, technology, science, society, market and local requirements. Nelson and Winter's theory aims to avoid generalisations from one industry to another or from one firm to another since it holds that each firm is normally committed to a specific technology and operates in a specific context. The search for new solutions to particular problems involves significant learning procedures as well as search processes based on R&D activities (Rothwell *et al.*, 1976; Cooper, 1980; Pavitt, 1987, Tidd *et al.*, 1997; Trott, 2002).

The organic approach leading to the fifth model of innovation

Another approach, known as the organic approach, was proposed by Clark and Juma (1987) in which the concept of interaction between science, technology, society, market and local requirements is amplified. This approach is based on the systems theory that unifies static structures and dynamic evolution in one general framework. This is why organisations conducting or facilitating innovation require time to gain capability, experience, knowledge and information in order to formulate change and adjust their conduct and policy. It is not only organisations which generate innovative activities, but also the 'network' between firms, subcontractors and government institutions. Clark and Juma argue that institutions provide financial support for the development of innovation and also shape its selection mechanism by providing the feedback mechanisms between external environments and technical development. The significance and influence of interactions and feedback mechanisms on the development of innovation are further explored by the recent literature on innovation in which the debate has shifted from an emphasis on internal structure to external linkages and processes (Tidd *et al.*, 2001). This demonstrates the greater importance being placed upon inter-organisation linkages and the structures of the public and private sector context within which innovation prospers (Rothwell, 1992; Cooke and Morgan, 1993; Marceau, 1996). Networks are increasingly being used to respond rapidly to fast-changing needs through the pooling of resources and the sharing of risk (Hobbay, 1996). This form of collaboration can assist in dealing with growing environmental uncertainty and complexity resulting from globalisation of markets and rapidly changing technologies (Granstrand *et al.*, 1992). It is also seen as a means to facilitate learning, transfer of technology and innovation (Dodgson, 1996; Leonard-Barton, 1990; Pieter van Dijk and Sandee, 2001).

The significance of the spatial or geographical dimension of innovation has been demonstrated by research aimed at understanding the successful development of innovation in specific locations such as the M4 corridor in the UK and Silicon Valley in California. This research has highlighted the importance of concepts such as industrial districts, innovative milieux and regional innovation networks (Porter, 1990; Camagni, 1991; Cooke and Morgan, 1993; Kanter, 1995; Morgan, 1997; Maillat and Kébir, 1998). The emergence of regional concentrations of innovation has reinforced the view that innovation is a collective process, dependent on many different interactions between an organisation and its external environment, which includes suppliers, customers, technical institutes, training bodies, technology transfer agencies, trade associations and other government agencies.

In line with the neo-Schumpeterien research, the current debate also supports the idea of adopting a multidisciplinary and integrated approach in trying to evaluate effectively the innovative potential of an organisation, an industry, a region or a nation. The value for such an integrated approach is that it implies that innovation should be seen not as a separate activity, but as a whole, integrating the organisation with its entire external environment (Grindey, 1993).

The neo-Schumpeterien research places great emphasis on areas which include the cumulative aspects of technology, the importance of incremental as well as radical innovations and the multiple inputs to innovation from diverse sources within and outside the firm. For the Neo-Schumpeterians, innovation is an evolutionary process which is cumulative, interactive, institutional and disequilibrating. It is a highly complex phenomenon where changes are of techno-economic and social types since they affect not only engineering activities but also institutions and various facets of management and organisation activities.

Many studies, both theoretical and empirical, have been carried out in order to identify the causes and the factors likely to generate innovation. However, although this research has significantly improved the understanding of the intricate relationships which cause innovation, there is still not a satisfactory explanation of the origin of innovations. The sources and the process of innovation are rarely confined within the boundaries of individual organisations. This explains the complex and uncertain aspect of innovation which requires the combination of inputs from a multiplicity of sources.

Models of innovation

There has been considerable debate about the causes of innovation. Early theories described innovation as a linear process comprising a succession of functional activities. Subsequent models viewed innovation as a coupling and matching activity marked by a multi-factor process which requires high levels of interaction and integration at intra- and inter-organisation levels.

Linear model: Technology-Push versus Need-Pull model

The dominant model of innovation has for a long time been the linear model, according to which innovation is a sequence of stages starting either from scientific research and known as 'Technology-Push' or some perception of a demand known as 'Need-Pull'. Schumpeter and Schmookler were among the first to recognise innovation as a primary engine of growth. For both of them, although in quite different ways, innovation was a very important component of economic development.

For Schumpeter, it is only by introducing radically new ideas into economic life that development can be generated. This model of innovation emphasised scientific and technological advance, and suggested that discoveries in basic science led to industrial technological development. Innovation in this model is based on a science-based technological change and is the outcome of professional Research and Development

(R&D) activities undertaken by qualified engineers and scientists. However, it implies a more or less passive role for the user (Rothwell, 1986). For the Need-Pull model, innovation arises in response to the recognition of a perceived market need. It focuses on the user or the customer as the starting point which triggers the whole process of innovation.

The debate about whether Technology-Push is distinguishable from Need-Pull and, if so, which is more important, was somewhat protracted and had no very clear outcome. Studies such as Johnston (1975) and Rothwell (1977) suggest that Need-Pull model is the most important determinant of the innovation process. Similarly, Langrish *et al.* (1972) and Gheorghiou *et al.* (1986) argue that very few innovations resulted from a scientific discovery.

For Rothwell (1992) and Freeman (1996), R&D plays an important role in ensuring that innovations are available when required. However, it is not the only generator of innovation. It is therefore too simplistic to opt for a straightforward model of classification because of the interaction which occurs within the innovative activity between the marketing and the R&D activities. Consequently, innovation can have a major significance in firms' marketing strategies as it can assist them in generating new markets and competitive advantage. There is an obvious link between a perceived need where marketing has a considerable role to play and technological development based essentially on R&D activities.

The coupling model: combining Technology-Push and Need-Pull models

Mowery and Rosenberg (1979) assert that both a knowledge base of science and technology and the market are important determinants of success in innovation. They discount any particular factor as the sole or the fundamental determinant of innovation and claim that the coupling of technology and the market is essential if an innovation is to be successful. This is why Pavitt (1984b, 1985), Rothwell (1992) and Dodgson (1996) regard R&D as a major component, but not the sole generator, of innovation. Innovation is increasingly viewed as the result of a conjunction between a perceived need and technological development. In their study of innovations, Gheorghiou *et al.* (1986) found that very few innovations fitted these models in a clear and unambiguous way. Innovation was caused by a combination of factors rather than a single one. This 'single factors' explanation of the innovation process was also rejected by the SAPPHO project (1974) This multi-factor approach is convincingly supported by Tidd *et al.* (1997) for whom most of the major innovations take place as a result of the interaction of technology, science and market.

Integrated model

The above discussion has shown that innovation is very complex and cannot be seen as being caused by a single factor. This argument goes far beyond the crude dichotomies discussed above on Technology-Push versus Need-Pull theories. This growing understanding of the process of innovation has been advanced by works such as that of Freeman *et al.* (1982), Nelson and Winter (1982), Dosi (1982, 1984), Clark and Juma (1987), Freeman and Perez (1988), Rothwell (1992, 1996), Nelson (1992), Freeman (1996), and Tidd et *al.* (1997).

As already discussed, Nelson and Winter (1982) explain the innovation process in terms of the interactions between an organisation's natural trajectory and the selection environment. The ability of an organisation to innovate is determined by market environments which are viewed as the basis of natural selection. Thus, in addition to the interaction of Technology-Push and Need-Pull, Nelson and Winter's theory introduces the concepts of natural trajectory and selection environment. They further explain that differences in innovative capabilities of organisations largely arise from differences in their institutional behaviour and strategy.

Similarly, Dosi (1982) argues that innovation is a cumulative process of iteration between technical feasibilities and market possibilities. For Freeman and Perez (1988), the interaction is not limited merely to market and technology but also affects the economic, social and institutional context in order to determine the best practice pattern for innovation. This concept of interaction has also been expanded by Clark and Juma (1987) who, in their case study, suggest that innovation is the outcome of the cumulative scientific and technical know-how and the organisations' expenditures on R&D. As previously explained, these authors assert that innovation also depends on feedback mechanisms between external environments and technical developments provided by institutions.

Systems integration and networking model

Rothwell (1992) claims in his historical study of technological change that the present model of innovation is significantly and increasingly being influenced by the formation of networks, collaboration and alliances leading to a variety of external relationships. He calls this model a 'fifth generation model' marked by systems integration and networking.

The review of the different models, as illustrated in Fig. 4.4 and Table 4.1, has contributed substantially to our understanding of innovation. They provide a more realistic and accurate representation of the innovation process, which can help us to implement and effectively manage innovation.

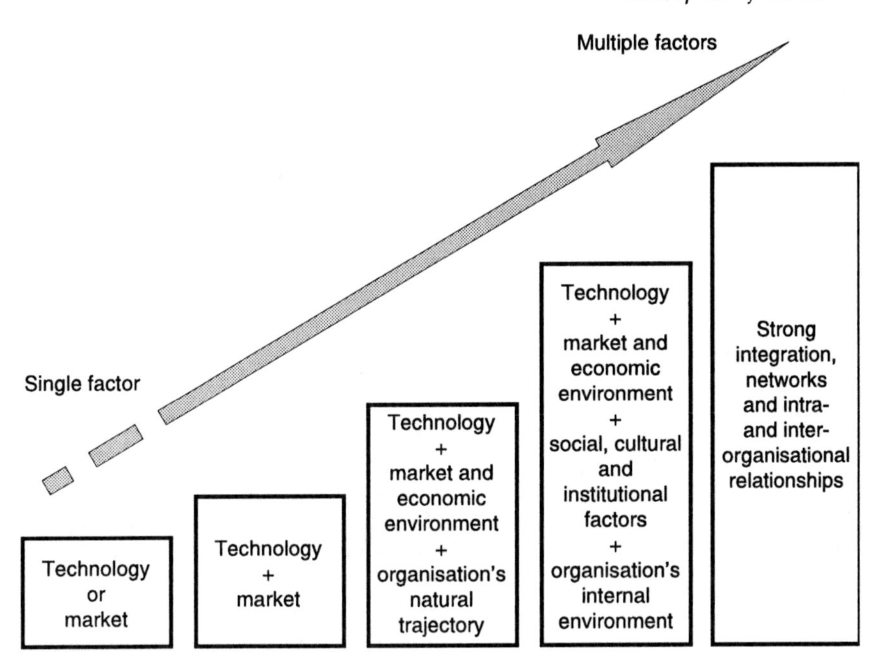

Fig. 4.4. Progression from single to multiple factor analysis (source: Saad, 1991 and 2000)

Process of innovation

The complexity of innovation makes any attempt to describe this process very difficult. However, the management of innovation requires a thorough understanding of the main stages through which an innovation is developed. This is why most studies of innovation have utilised the concept of activities occurring in stages.

The concept of stages in the process of innovation

Zaltman *et al.* (1973) explain that innovation is composed of a set of stages. For Cooper (1980), stages consist of decisions, actions and related behaviours taken at different times which, in a logical sequence, move the process towards an outcome. This concept of stages is useful to organise and manage the process of innovation which is considered as being long, complex, dynamic and showing interaction and overlap between its elements. It is, however, important to emphasise that dividing the innovation process into discrete stages represents a rationalisation rather than reality, particularly if the progression through the stages is supposed to be linear. This implies that each stage will be clearly

Table 4.1. Main models of innovation

Models	Authors	Major contributions
Technology-Push	Schumpeter	Economic growth is achieved by introduction of a new idea where science and technology are the major sources of innovation
Clusters of innovation	Mensch	Stalemate creates an accelerator mechanism and induces innovation which comes in clusters
Need-Pull	Schmookler	Innovation is the result of emphasis put on demand factors
Coupling model	Mowery and Rosenberg	Technology and demand are both determinants of success in innovation
Long cycles of the world economy	C. Freeman	Electronics industries considered to form the basis of a fourth Kondratiev wave with innovation arising in the upswing phase as an outcome of both market and technology
Natural trajectories	Nelson and Winter	Innovation is viewed as an interaction between the firm's natural trajectory and the selection environment
Technological paradigm	Dosi	Technological paradigm incorporates interrelationships between scientific progress, technical change and economic development and suggests a continuous progress along a defined technological trajectory
Social and economic paradigm	Freeman and Perez	Innovation is viewed as an economic interaction between the economic, social and institutional spheres
Organic	Clark and Juma	Innovation is an evolutionary process cumulative through time within a social system where institutions provide feedback mechanisms between external environments and technical development
Regional network paradigm	Porter, Kanter, Camagni, Cooke and Morgan	Significant links between innovation and regional support and learning
Systems integration and network paradigm	Rothwell	Innovation as a multi-factor process depending on intra- and inter-organisational relationships

Source: Saad (1991 and 2000).

identifiable whereas, in practice, the boundaries between stages are usually blurred, and their order may vary. It also assumes that the process has a clear start and that activities are equally distributed among the different stages. It is clear that such assumptions are strongly rejected by the literature. Cooper finds it difficult to generalise about the order in which the process works and argues that linear progression in innovation contradicts historical facts. As Rosegger (1980) argues, the linear process does not take into account the numerous and complicated feed-back mechanisms influencing the process of innovation. In practice, and as illustrated by studies such as Lambright (1980), the linear progression model fails to account for the way in which decisions affect each other within the process of innovation.

As a solution, Rosegger (1980) proposes a model which is summarised in Fig. 4.5 and which is based not on stages, but on a multi-cycle search for information in which decision-makers can return a new idea to a previous cycle, terminate it at any level of development, speed it up or slow it down depending on the existing circumstances. They can also modify this idea in response to feedback on changes from both the internal and external environment.

This representation addresses innovation mainly from the perspective of the producer of the innovation. However, transfer and implementation

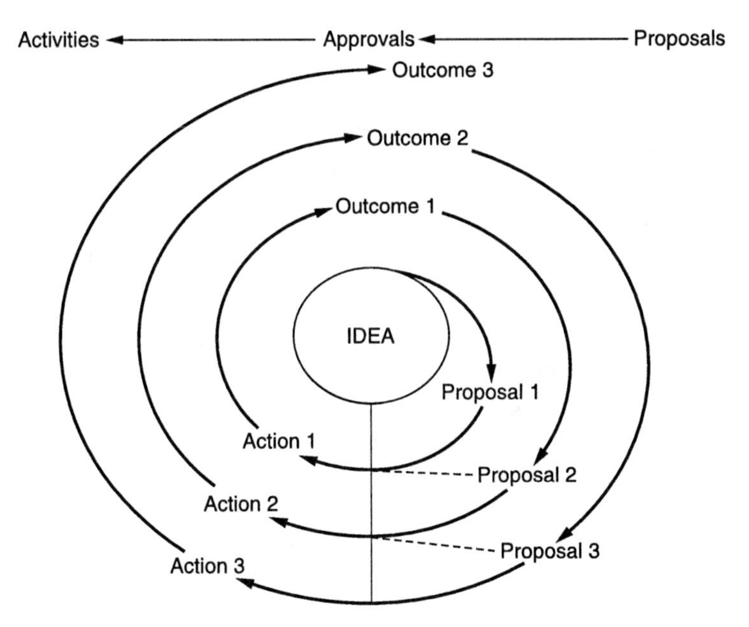

Fig. 4.5. Rosegger's model of innovation (derived from Rosegger, G. The Economics of Production and Innovation)

from one sector of the economy to another is not acknowledged in this model. Another model representing and describing innovation from the user's point of view is proposed in Fig. 4.6.

Key stages in the process of innovation

Although there are risks in oversimplifying the process of innovation, it is nevertheless necessary to adopt a systematic approach or model to describe the main stages of this process. The proposed model is divided into five major stages (Fig. 4.7) namely:

- identification of the need to innovate
- knowledge awareness
- choice
- planning and
- implementation.

The rationale for these five stages is based on the crucial need to acquire an in-depth understanding of the mechanisms by which key stages of innovation can effectively be shaped and managed.

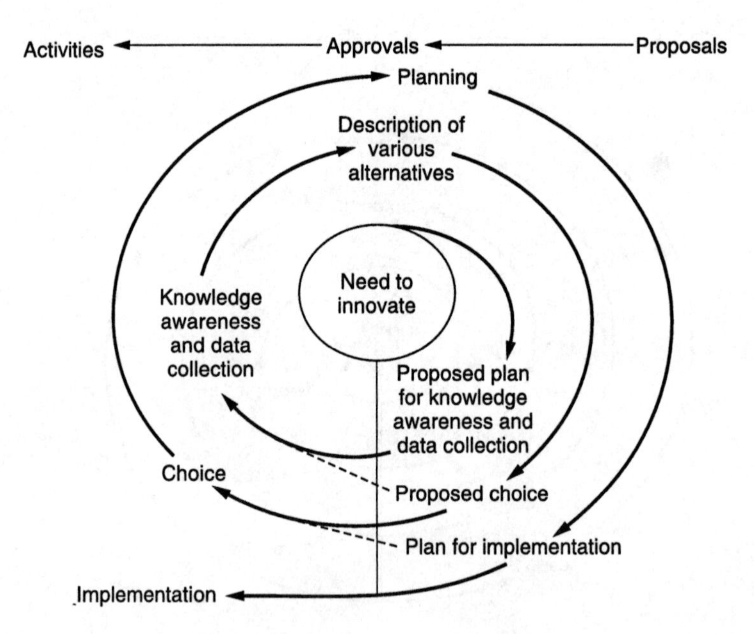

Fig. 4.6. Models of innovation from the user's persepective (source: Saad 1991 and 2000)

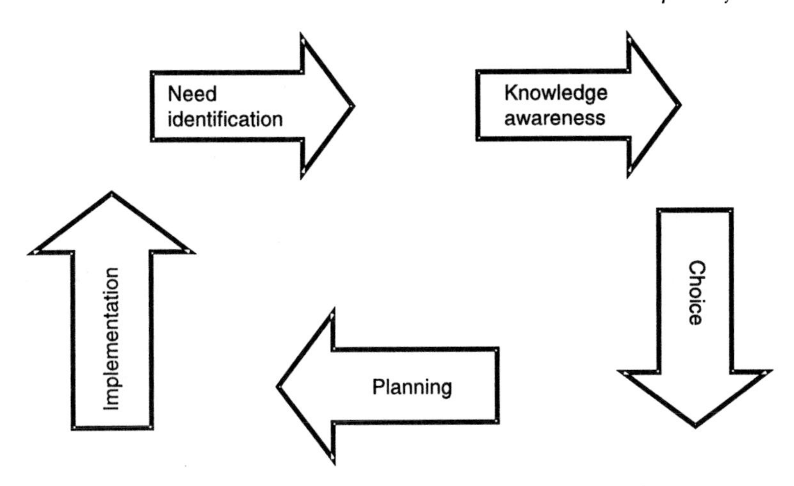

Fig. 4.7. Main stages of the process of innovation process (source: Saad 1991 and 2000)

Stage 1: identification of the need to innovate. In this stage, the need for change is identified and defined in order to minimise the risk of making an incorrect decision. As already mentioned, the debate on the different models of innovation clearly shows that the need to innovate rarely occurs as a result of a single motive but rather as an aftermath of diverse concurrent motives. This requires strong interaction and feed-back mechanisms between internal and external factors such as in-house capabilities and economic and social factors.

Stage 2: developing awareness. The aim of this stage is to gain a greater understanding and awareness of the new idea with the view of minimising difficulties of implementation and ensuring success. It starts when the first information on a new idea enters the organisation. The objective is to acquire information, knowledge and understanding of the key features and the functioning principles of the new idea and, their appropriateness to meeting the need to innovate. Different forms of information gathering and learning are needed to reach the required level of awareness. If an adequate level of information and understanding are not gained, other stages are likely to be adversely affected. An incorrect decision will lead to an incompatibility between the new idea and the need to innovate as identified by the organisation.

Stage 3: selecting the innovation. Subsequent systematic evaluation then leads to a decision to adopt or reject a new process or idea, having fully considered all problems likely to occur during the implementation stage. In this stage, the major aim is to collect the data necessary to explore the strengths and the weaknesses of the innovation in order to determine how best it can be successfully introduced into a given situation. Alternative innovations are listed, evaluated and compared on the basis of their main characteristics. From a review of studies of the influential characteristics of innovation (Zaltman *et al.*, 1973; Nabseth and Ray, 1974; Bessant, 1982; Rogers, 1971, 1983), the key attributes to examine prior to any selection or adoption of innovation are:

- relative advantage
- compatibility
- complexity and
- key individuals or champions.

Rogers defines relative advantage as the degree to which an innovation is perceived as better than the idea it supersedes. The degree of relative advantage is measured in economic terms such as reduction of production cost, productivity increase, saving in time and effort and immediacy of the reward. The greater the perceived relative advantage of an innovation, the more rapid and successful its rate of adoption. For Rogers and Shoemaker (1971), compatibility is the degree to which an innovation is perceived as consistent with existing values, past experiences and needs of the receivers. It concerns the 'fit' between innovation and the context or the environment in which it is going to be implemented. Rogers (1983) views complexity as the degree to which an innovation is perceived as difficult to understand and use. A new idea which is simple to understand will be adopted rapidly. A more complex innovation which requires new skills and understanding may lead to complex implementation and increased risk and uncertainty. Key individuals are those responsible for identifying and disseminating information about innovation opportunities.

The outcome of the selection stage is the decision to adopt an innovation which is seen as the most appropriate. However, this is not yet the end of the innovation process as this idea has to be implemented. Adoption is only the final step of the conceptual activity based on information gathering and processing and awareness raising.

Stage four: planning. This stage is needed for an effective transition between the conceptual activities identified in Stages 1, 2 and 3 and in the implementation stage the innovation is put into practice. The main objective of the planning stage is to anticipate events which are likely to occur and to ensure the best fit between the selected innovation and the context in which it is going to operate.

Stage five: implementing. This stage is increasingly seen as being at the heart of successful innovation (Leonard-Barton, 1990; Voss, 1991). Implementing an innovation is getting it 'up and running' in daily operations. The process of implementation needs to be linked to the organisation's background and culture in order to ensure compatibility and success. It is a prolonged process entailing progressive development, evaluation, adaptation and modification. Consequently, implementation is not simply about adopting an innovation as advocated by the classical diffusion literature (Rogers and Shoemaker, 1971). It is a continuous development process putting decisions into actions with a significant reflection of the very considerable obstacles that may have to be overcome.

The importance of on-going change featuring implementation is also highlighted by Leonard-Barton (1988) who depicts this stage as a dynamic process of mutual adaptation, between the innovation and its environment. Effective implementation is also significantly associated with the implementation 'champion', incremental stepwise installation, workforce engagement, training and cross-functional implementation teams (Leonard-Barton and Deschamps, 1988; Leonard-Barton, 1988 and 1990; Winch and Voss, 1991; Voss, 1985; 1991 and 1992). Empirical studies have shown that in spite of all the conceptual work completed in the previous stages to reduce uncertainty and the risk of failure, risks associated with implementation cannot be entirely eliminated.

At this stage, and as a result of learning and experience, the innovation must be reviewed and modified in order to match it to the changing

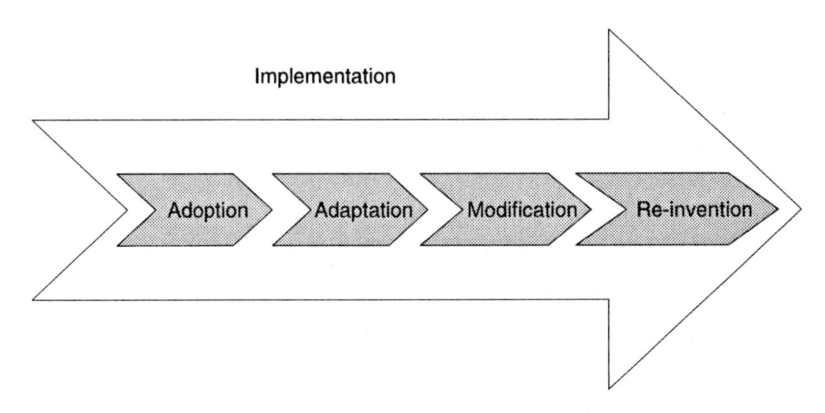

Fig. 4.8. Implementation stage (source: Saad 2000)

requirements of the organisation and its environment. These activities have been variously termed: routinisation, incorporation, stabilisation or continuation (Lambright, 1980; Zaltman *et al.*, 1973). This stage includes adaptation, modification and re-invention.

The adaptation activity brings about changes to the innovation to ensure success and reduce any incompatibility with the environment. Achieving compatibility may also require changes within the organisation. Indeed van de Ven, quoted by Leonard-Barton (1988), claims that innovations not only adapt to existing organisational and industrial arrangements, but they also transform the structure and practice of these environments. Modification involves the development activities, training and learning that the organisation undertakes in order to improve and sustain the performance of the implemented innovation.

Re-invention is the activity whose sole purpose is to renew the innovation according to the context in which it operates. It is described by Rice and Rogers (1980) as an alteration of the original innovation to suit users' needs. For Johnson and Rice (1987) this phase is important when the innovation is used in ways unforeseen by the original developers. At this step, the organisation is likely to be in possession of a level of learning and experience which encourages them to undertake substantial changes to the selected and implemented innovation.

Innovation is a lengthy, complex and dynamic process which suggests the need for a systematic management approach involving key stages such as knowing, understanding, planning, implementing and sustaining the new idea.

Key determinants in managing innovation

The success or failure of innovation is influenced by a whole range of factors which often vary from one organisation to another, from one industry to another, and even from one country to another. Several key attributes of successful and unsuccessful innovations have been identified by empirical work. Different studies highlight different factors, but success is significantly associated with combinations of particular factors (Swords-Isherwood, 1984; Tidd *et al.*, 1997 and 2001; Trott, 2002).

One of the earliest systematic studies of factors involved in the initiation of commercial innovation was carried out by Carter and Williams (1957). This study, which was essentially focused on factors related to success, identified that innovation was impeded by the paucity of marketing. Teubal *et al.* (1976) in their FIP study (Falk Innovation Project) also observed the lack of market appraisal and marketing research in British organisations and their impact on innovation. Myers and Marquis (1969) in their study suggested that successful innovation is not the result of a single action but rather a total process. They also identified that innovation arising from market factors (Need-pull model) was more frequent in small companies while innovation arising from technical or production factors (Technology-push model) was more frequent in large companies. Overall, however, recognition of demand was seen as a more frequent factor in innovation than recognition of technical potential.

Langrish *et al.* (1972) investigated 84 firms, all of which gained the Queen's Award to Industry for innovation between 1966 and 1969. They identified the main factors of success as being a clear identification of a need, top management commitment, good cooperation, availability of resources and support from government. On the other hand, the lack of marketing activity, poor communication, shortage of resources and resistance to new ideas were identified as factors inhibiting innovation.

The SAPPHO (1974) study led by the Science Policy Research Unit of Sussex University, examined successful and unsuccessful innovation in the chemical and scientific instrument industries. The findings of this project confirmed that the use of marketing and the identification of user needs (Need-pull model) are significantly linked to the success or failure of innovation. This project also suggests that success requires competence, experience and effective communication. Failure, on the other hand, was associated with a lack of awareness of user requirements, effective management, communication and market research. Hence, their rejection of a single factor explanation of innovation.

The importance of R&D activities has been established by studies such as those conducted by Freeman (1996) who identified that the factors of success include strong in-house professional R&D as well as performance of basic research, readiness to take risks, identification of a need and market research, and effective internal and external communication. More recent research (Camagni, 1991; Cooke and Morgan, 1993; Kanter, 1995; Porter, 1996; Morgan, 1997; Tidd *et al.*, 2001) highlighted the importance of factors such as regional support and networks for learning and innovation.

These studies suggest a number of common factors characterising successful innovations. However, drawing such general conclusions does not always correspond to reality. As already discussed, each organisation has its own natural trajectory and selection environment and it is quite risky to generalise about determinants of innovation which are contingent on their environment, location and national systems of innovation (Lundvall, 1990; Cooke and Morgan, 1993; Nelson, 1993; Porter, 1996).

This review of the innovation process and the different models of innovation helps to give an insight into construction's difficulty in innovating. These difficulties can be explained by the lack of recognition of a number of key determinants of innovation as outlined below:

- identification of a clear need for change
- responsiveness to internal and external change
- the achievement of good linkages within and between organisations leading to more collaborative relationships
- treating innovation as a corporate-wide task
- adopting a strategic approach in the management of innovation
- developing and sustaining a supporting organisational culture for innovation
- top management commitment and acceptance of risk
- presence of certain key individuals or champions
- effective and on-going learning process
- systematic approach to developing, implementing, monitoring and sustaining innovation
- external support.

Consequently, success in innovation, as investigated in the remaining part of the book, is rarely associated with doing one or two things outstandingly well. It is dependent on performing all functions competently and in a well balanced and integrated manner, taking into account the specificities of internal and external environment.

References

BESSANT, J. *Influential Factors in Manufacturing Innovation Research Policy*, **11**, pp. 117–132, 1982.

BESSANT, J. *Fifth Wave Manufacturing: Management Implications of Advanced Manufacturing Technology*. Blackwell Publishing, Oxford, 1991.

BURNS, T. and STALKER, G. *The Management of Information*. Tavistock, London, 1966.

CAMAGNI, R. Introduction: From the Local 'Milieu' to Innovation Through Co-operation Networks. *Innovation Networks: Spatial Perspectives*. CAMAGNI, R. (ed), Belhaven Press, London, 1991.

CARTER, C. F. and WILLIAMS, B. R. *Industrial and Technical Progress*. Oxford University Press, 1957.

CLARK, N. and JUMA, C. *Long-Run Economics: An Evolutionary Approach to Economic Growth*. Frances Pinter Publishers, London/New York, 1987.

COOKE, P., and MORGAN, K. The Network Paradigm: New Departures in Corporate and Regional Development. *Regional Science*, **74**, No. 4, 317–340, 1993.

COOPER, C. *Policy intervention for technology innovation in developing countries*. World Bank. Staff working Paper. No 441. Washington DC, 1980.

DODGSON, M., Technological collaboration. *Handbook of Industrial Innovation*. DODGSON, M. and ROTHWELL, R. (eds), Edward Elgar, 1996.

DOSI, G. Technological paradigms and technological trajectories. *Research Policy*, **11**, No. 3, 147–162, 1982.

DOSI, G. *Technological Change and Industrial Transformation: The Theory and Application to the Semiconductor Industry*. MacMillan, London, 1984.

FREEMAN, C., CLARK, J. and SOETE, L. *Unemployment and Technical Innovation: A Study of Long Waves and Economic Development*. Frances Pinter, London, 1982.

FREEMAN, C. and PEREZ, C. Structural crises of adjustment, business cycles and investment behaviour in DOSI, G. *et al.* (eds), *Technical Change and Economic Theory*. Frances Pinter, London, 1988.

FREEMAN, C. Innovation and growth in *Handbook of Industrial Innovation*. DODGSON, M. and ROTHWELL, R. (eds), Edward Elgar, 1996.

GHEORGHIOU, L. *Post-Innovation Performance: Technological Development and Competition*. Macmillan, London, 1986.

GRANSTRAND, O., HAKANSON, L. and SJOLANDERS, S. (eds) *Technology Management and International Business*. John Wiley & Sons, Chichester, 1992.

GRINDLEY, P. Firm-strategy and successful technological change. *Handbook of Innovation Management*. COZIJNSEN and VRAKKING (eds), Edward Elgar Publishing, 1993.

HOBBAY, M. Innovation in semiconductor technology; the limits of the Silicon Valley network model. *Handbook of Industrial Innovation*. DODGSON, M. and ROTHWELL, R. (eds.), Edward Elgar, 1996.

JOHNSON, B. and RICE, R. *Managing Organisational Innovation: The Evolution from Word Processing to Office Information Systems*. Columbia University Press, New York, 1987.

JOHNSTON, P. S. *The Economics of Invention and Innovation, with a Case Study of the Development of the Hovercraft in the UK*. Martin Robinson & Co Ltd., London, 1975.

KANTER, R. *World Class: Thriving Locally in the Global Economy*. Simon and Schuster, New York, 1995.

KENDRICK, J. W. *Improving Company Productivity; Handbook with Case Studies*. The John Hopkins University Press, Baltimore, 1984.

KUHN, T. *The Structure of Scientific Revolutions*. Harvard University Press, 1962.

KUZNETS, S. *Economic Growth and Structure: Selected Essays*. W. W. Norton and Co. Inc., New York, 1965.

LAMBRIGHT, W. H. *Technology Transfer to Cities*. Westview Press, Boulder Co., 1980.

LANGRISH, J. *et al. Wealth from Knowledge*. Macmillan, London, 1972.

LAWRENCE, P. and LORSH, J. *Organisation and Environment*. Harvard University Press, 1967.

LEONARD-BARTON, D. Implementation and mutual adaptation of technology and organisation. *Research Policy*, **17**, 1988, 251–277.

LEONARD-BARTON, D. and DESCHAMPS, I. Managerial influence in the implementation of new technology. *Management Science*, **34**, No. 10, October 1988.

LEONARD-BARTON, D. Modes of technology transfer within organisations: point-to-point versus definition. *Production and Operations Management*. Harvard Business School, May, 1990.

LEONARD-BARTON, D. *The Factors as Learning Laboratory*. Harvard Business School Working Paper, No 92–023, 1991.

LUNDVALL, B. *National Systems of Innovation: Towards a Theory of Innovation and Interactive Learning*. Frances Pinter, London, 1990.

MAILLAT, D. and KÉBIR, L., Learning region, milieu innovateur et apprentissages collectifs. *Le paradigme de milieu innovateur dans l'économie spatiale contemporaine*. Atti del colloquio GREMI, Parigi, 1998.

MARCEAU, J. Another determinant – clusters, chains and complexes: three approaches to innovation with a public policy perspective. *Handbook of Industrial Innovation*. DODGSON, M. and ROTHWELL, R (eds), Edward Elgar, 1996.

MENSH, G. *Stalemate in Technology: Innovations Overcome the Depression*. Ballinger Publishing Co., Cambridge, Mass, 1979.

MORGAN, K. The learning region; institutions, innovation and regional renewal. *Regional Studies*, No. 31.5, 491–503, 1997.

MOWERY, D. and ROSENBERG, N. The influence of market demand upon innovation: a critical review of some recent empirical studies. *Research Policy*, **8**, 1979.

MYERS, S. and MARQUIS, D. G. *Successful Industrial Innovation*. National Science Foundation, Washington DC, 1969.

NABSETH, L. and RAY, G. *The diffusion of New Industrial Processes*. Cambridge, Cambridge University Press, 1974.

NELSON, R. and WINTER, S. *In Search of a Useful Theory of Innovation*. Research Policy 6, 1977, pp. 36–76.

NELSON, R. and WINTER, S. Innovation and economic development: theoritical; retrospect and prospect. *IDB/CEPAL Studies on Technology and Development in Latin America*. 1978.

NELSON, R. and WINTER, S. *An Evolutionary Theory of Economic Change*. 1982. Belknap Press, Cambridge, Mass. and London.

NELSON, R. (ed). *National Innovation Systems*, Oxford University Press, Oxford, 1993.

PAVITT, K. *Technology, Innovation and Strategic Management*. Brighton Science Policy Research Unit, University of Sussex, Mimeo, 1984a.

PAVITT, K. International patterns of technological accumulation. In HOOD, N. and VAHLNC, J. E. (eds), *Strategies in global Competition*, Croom Helm, London, 1988.

PAVITT, K. *Patent Statistics as Indicators of Innovative Activities, Possibilities and Problems*. Scientometrics, 1985, 7, 77–99.

PAVITT, K. On The Nature of Technology. Inaugural Lecture, Science Policy Research Unit, Sussex University, 23 June 1987.

PIETER VAN DIJK, M. and SANDEE, H. (eds). *Innovation and Small Enterprises in the Third World, New Horizons in the Economics of Innovation*. Edward Elgar, 2002.

PORTER, M. *The Competitive Advantage of Nations*. Macmillan, London, 1990.

PORTER, M. Competitive advantage, agglomeration economies, and regional policy. *International Regional Science Review*, **19**, 1 & 2, 1996, 85–94.

RICE, R. E. and ROGERS, E. M. Re-invention in the innovation process. *Knowledge*, **1**, 499–514, 1980.

RICKARDS, T. *Stimulating Innovation: A System Approach.* Frances Pinter, London, 1985.

ROGERS, E. and SHOEMAKER, F. *The Communication of Innovations.* Free Press, New York, 1971.

ROGERS, E. *The Diffusion of Innovations* 3rd edn. Free Press/MacMillan, New York, 1983.

ROSEGGER, G. *The Economics of Production and Innovation. An Industrial Perspective.* Pergamon Press, London, 1980.

ROSENBERG, N. *Learning by Using.* Stanford University, 1979.

ROSENBERG, N. *Inside The Black Box: Technology and Economics.* Cambridge University Press, Cambridge, 1982.

ROTHWELL, R. The characteristics of successful innovators and technically progressive firms. *R&D Management*, **7**, 1977, 191–206.

ROTHWELL, R., *et al. Methodological Aspects of Innovation Research – Lessons from a Comparison of Project SAPPHO and FIP.* Science Policy Research Unit, Sussex University, 1976.

ROTHWELL, R. Innovation and re-innovation: a role for the user. *Journal of Marketing Management*, No. 2, 1986, 109–123.

ROTHWELL, R. Successful industrial innovation: critical success factors for the 1990s. *R&D Management*, 22 (3), 1992, 221–239.

RUTTAN, V. Usher and Schumpeter on invention, innovation and technological change. *Quarterly Journal of Economics*, November, 1959, 596–606.

SAAD, M. The transfer and management of new technology: The case of two firms in Algeria, unpublished PhD thesis, Brighton Polytechnic, 1991.

SAAD, M., Development through technology transfer – creating new organisational and cultural understanding. *Intellect*, 2000.

SAPPHO. *Success and Failure in Industrial Innovation.* Report on SAPPHO project by Science and Policy Research Unit of Sussex University, 1974.

SCHON, D. A. *Technology and Change: The New Heraclitus.* Delacorte Press, New York, 1967.

SCHUMPETER, J. A. The Instability of Capitalism. *Economic Journal*, 1928, pp. 361–386.

SCHUMPETER, J. A. *The Theory of Economic Development.* Harvard University Press, Cambridge, Mass., 1934.

SCHUMPETER, J. A. *Business Cycles: A Theoritical, Historical and Statistical Analysis of the Capitalism Process.* Vol. 1, McGraw-Hill, New York, 1939.

SOLOW, R. Technical change and the aggregate production function. *Review of Economics and Statistics*, **39**, 1957.

STRASSMAN, W. P. *Risk and Technological Innovation: American Manufacturing Methods during the 19th Century.* Cornell V.P., 1959.

SWORDS-ISHERWOOD, N. *The Process of Innovation: A study of Companies in Canada, the United States and the United Kingdom.* British–North American Committee, John Wiley & Sons, Chichester, London, 1984.

TEUBAL, M. *et al. The Falk Innovation Project (FIP).* 1976.

TIDD, J. *Integrating Technological Market and Organisational Change.* Wiley and Sons, Chichester, 1997.

TIDD, J., BESSANT, J. and PAVITT, K. *Managing Innovation*: 2nd edn, John Wiley & Sons, Chichester, Chichester, 2001.

TRIST, E. *The Evolution of Socio-Technical Systems; A Conceptual Framework and an Action Research Programme*. Occasional Paper No. 2, Ontario QWL, 1981.

TROTT, P. *Innovation Management and New Product Development*, 2nd edn, Financial Times – Prentice Hall, 2002.

USHER, A. *A History of Mechanical Invention*. Harvard University Press, Cambridge, Mass., 1954.

VOSS, C. A. *The Need for a Theory of Implementation of Innovation*. ESRC Conference, Cumberland House, May, 1985.

VOSS, C., A. *The Process of Implementation of New Processes*. Business Strategy Review, 1991.

VOSS, C. A. *Successful Innovation and Implementation of New Processes*. Business Strategy Review, 1992.

WINCH, G. and VOSS, C. A. *The Process of Implementation, The Evaluation Stage, Operations Management*. Working Paper, London Business School, 1991.

WOODWARD, J. *Industrial Organisation*. Oxford University Press, Oxford, 1965.

ZALTMAN, G. *Innovations and Organisations*. John Wiley, New York, 1973.

Chapter 5

The effectiveness of project management in implementing innovation in construction

Project management is the main instrument in unifying the diverse range of participants and procedures in the design and construction process. Consequently, it remains the principal vehicle for implementing and sustaining innovation in most of the construction sector. Implementation, which is at the heart of the innovation process, is about effecting changes in day-to-day operations. The project is the most readily available context within which innovation is implemented and where key participants can be influenced and supported in their attempts to change in order to improve relationships, integrate processes and increase customer focus. The project can also be seen as a main vehicle for shared learning and the diffusion of innovation within the project team and throughout the industry.

This chapter examines the evolution of project management during the past 60 years or so in order to investigate how it has implemented and sustained innovation in responding to changes in the paradigm. The first part investigates the influence of Fordism on project management thinking. The second section assesses how it is responding to the emergence of post-Fordism. The third section reviews the progress of project management in construction towards the adoption of the new paradigm and questions its effectiveness as a tool for innovation with reference to the conceptual framework for innovation identified in Chapter 4.

Critical review of project management and its strong Fordist roots

The basic purpose for initiating a project is to accomplish some beneficial objectives and goals. Project management is essentially an organisational innovation, which involves organising the task to be completed as a project in order to focus the responsibility and authority for the attainment of the project goals on an individual or small team. The project manager is expected to coordinate and integrate all activities needed to reach the project's goals. The concept of projects and their management has gained widespread usage in a diverse range of activities including building and civil engineering, power, petrochemicals, the extractive industries of oil and gas, pharmaceuticals, shipbuilding, aerospace, information systems, telecommunications and defence. Recently, the concept of projects has changed and it has ceased to be dominated by project-based industries. There are other activities which, although not necessarily organised primarily on a project basis, are increasingly adopting some form of project-based management. These include manufacturing, emergency and social services, entertainment, education, consulting and financial services.

Although a wide variety of projects have been managed since the start of civilisation, the discipline of project management is a phenomenon of the twentieth century and, therefore, greatly influenced by Fordist and post-Fordist thinking. It began to emerge in the 1920s in the US aircraft industry when the US Air Corps' Material Division moved progressively towards a project office function to monitor the design, development and production of aircraft. It was further developed in the US Defence programme of the 1950s and its space programme in the 1960s, where it was closely related to systems engineering. At the same time, Exxon and other comparatively young process engineering companies also began to develop a Project Engineer function. This allowed an engineer to progress a project through its various functional departments with the aim of achieving greater coordination between the various functions and groups.

The growing complexity of projects in the early years of the twentieth century increased the realisation of the need to develop more considered, consistent and scientific approaches to work organisation and undertaking complex operations. The early theories of management to replace the ad hoc, rule of thumb approach to organisations – Weber's work on bureaucracy and Taylor, Gilbreth and Gantts's thinking on scientific management (as discussed in Chapter 2), significantly influenced the fledgling discipline of project management. The Scientific School of management sought to see organisations as rational and scientific entities characterised by the horizontal and hierarchical division of labour, and the minimisation of human skills and discretion. A major aspect of this rational approach was the development of planning techniques for production scheduling such as the Gantt bar chart. This technique, which was developed by Henry Gantt in the US in 1917, went on to become a major tool in the management of projects. The Theory of Work Harmonisation by Adamiecki around 1896 was the forerunner of work-flow network planning which became very popular with Critical Path Management (CPM) and Programme Evaluation and Review Technique (PERT). Path analysis was developed in 1918 by Wright as a way of decomposing relationships and using statistics to express their causal intensity.

The emerging essence of project management cut across, and in a sense conflicted with, the principles of the Scientific School and its approach to work organisation. The development of project management can be seen as challenging the prevailing pattern of organisation (almost invariably pyramidal or functional), which was increasingly being seen as alienating workers and not being conducive to horizontal integration. Gulick (1937) proposed the matrix organisation where a coordinator might be appointed to integrate the administration of a task involving several functional areas from within the organisation. The Air Force Joint Project Offices and Weapons Systems Project Offices in the US first implemented this attempt at greater integration of the different functions in the early 1950s. The appearance of the horizontal, or task, form of organisation in the academic writings on management began to challenge the pure functional structure of organisation associated with the scientific approach to management, where all staff were clearly and unambiguously in their functional groups.

In the 1950s, further developments in project management took place essentially in the process, construction, heavy engineering and defence sectors. The earliest obvious development of Program and Project Management began in the early 1950s in the US Air Force. The Cold War meant there was an urgent need to develop and produce large numbers of increasingly sophisticated aircraft and later long-range missiles. This pressure for change resulted in a number of new systems and techniques including the development of systems management and numerical techniques, which drew heavily on the discipline of Operational Research (OR). This involved the mathematical modelling of complex situations, a focus that still persists in project management today and which partly explains the strong rigidity and lack of focus on people which is associated with many project management approaches.

Project management has a number of roots, as well as those in the defence industries in the US. The oil, gas and petrochemical industries continued to emphasise the integrated approach to project identification and management. Morris (1994) refers to a technique, 'the controlled sequence duration' for plant maintenance scheduling, which was developed by ICI at their Billingham works in 1995. In a similar approach in 1957, the CEGB developed a technique to identify the longest irreducible sequence of events in the overhaul of generating plants – later termed the major sequence. Morris argues that both of these innovations were direct precursors of what soon became known as the critical path. However, neither were widely publicised and consequently had little impact on general practice, which demonstrates the limited and slow dissemination of ideas between different groups and industries.

Meanwhile, in the US, early computer technologies were beginning to impact on project management. Du Pont developed uses for its newly acquired Univac computer, including optimising the trade-off between the time and the cost of plant overhaul, maintenance and construction. In particular they used it to calculate activity durations with some reliability and hence determine the optimum duration for a large project. By 1957, a pilot scheme had been developed and run using the arrow diagram method (Morris, 1994).

The continuing emphasis on detailed and systematic control in the management of projects was illustrated by the development of formal tools and techniques to help manage large, complex projects characterised by uncertainty and high levels of risk. The RAND Corporation, a defence contractor, developed the Programme Evaluation and Review Technique (PERT) for planning the development of missiles. Du Pont, the chemical manufacturer, developed Critical Path Analysis (CPA) for scheduling the shut-down of its plants for maintenance. CPA and CPM formed the basis of project management systems over the next 20 years or so and came to dominate project management thinking. These methods incorporated a number of criteria for a project control system: a careful time estimate for each activity, no matter how far into the future; a recognition that estimates involve uncertainty and that each activity should have a probability distribution of the time it might require, and precise knowledge of the sequencing required or planned in the performance of activities. They also involved the collection of data necessary to accomplish tasks, the mathematical formula for determining the expected time of achieving the event, and the identification of the critical path – the sequence of events in the project that required the longest time for completion. These methods have survived and continue to dominate much of the current thinking in relation to project management. However, a fundamental weakness of these essentially numerical methods is that they are aimed at simplifying and transforming very complex situations into models following a linear progression and promoting the idea of 'one best way' for the management of projects.

Morris (1994) argues that CPM and PERT are strikingly similar although neither group became aware of the other's work until 1959. Du Pont's business in construction was fundamentally different from the Navy's in that technologies and processes were largely known, in contrast to the less certain and predictable R&D world of Polaris. Futhermore, the environment of the ballistic missile programmes was intensely schedule-driven, whereas Du Pont operated in a commercial environment in which costs mattered greatly. Its approach calculated various schedule durations according to varying resource utilisation. This allowed the costs of different resources to be calculated. CPM dealt not only with costs but also with resource allocation. In the defence systems development/procurement environment, uncertainties in technologies and processes reflected programmes that were largely R&D in nature before reaching their production phases. This led to one of the fundamental differences between PERT and CPM. PERT allowed estimators to bracket ranges of reasonably foreseeable time (and later cost) outcomes for each project task. Algorithms reduced these ranges to single-value ('expected') estimates. These were often substantially different, and less optimistic, than the traditional single-value estimates. Once combined through logic sequence of the activity network, the average values were re-expanded by algorithms to reflect the combined uncertainties of the variable activity estimates. Because Du Pont was cautious in promoting its technique, PERT received more attention, but by the late 1960s CPM had become the basic method of most construction scheduling, and was generally more commonly used than PERT.

During the late 1960s, a decade of experience with PERT, CPM and various derivatives had led to widespread acknowledgment in US defence procurement circles that forecasts of time (and cost) performance using activity network analysis were inherently optimistic and that even PERT had not fully addressed uncertainties and their interactions. By 1970, the Rand Corporation and individual OR teams supporting the US Navy, specifically the Project TRANSIM team led by Professor Alfred M. Feiler at UCLA, were using general purpose simulation tools to depict the interactive effects of uncertainties on projects, accommodating both variable estimates and variable logic (in future courses of events). Feiler's team became pioneers in applying these multi-iterative simulation techniques, not only within US defence procurement but also in hospital construction and other uncertainty-fraught capital project sectors. The application of project simulation formed the basis for the new discipline of project risk management, conceived and implemented as an enhancing accompaniment to conventional project planning, budgeting, monitoring and control. The discipline, from its inception, was heavily quantitative, having been born out of the need for more realistic and reliable information on which to base decisions that inevitably carried dimensions of time and cost.

In 1975, less than five years after Feiler's landmark applications in the US, Euro Log Ltd, assisted by Professor Feiler and his team, became Europe's first commercial practitioners for project risk management support. From its inception, London-based Euro Log introduced project risk management into the global oil and gas industry through offshore oil field developments. The company also became instrumental in 1984/85 in carrying out practical applications with the UK Ministry of Defence, and later participated in the ministry's major initiative in implementing project risk management as a standard procurement practice.

The 1990s saw a more positive emphasis on project risk management, recognising the equal importance of opportunity, its capture and exploitation with that of risk, its prevention and reduction. This has become an increasingly important trend with the migration of project risk management into investment risk management, where the project is viewed as a means to a business end. With this balancing of emphases, the management of project risk is evolving into the management of project uncertainty. The integration of project risk management with the management of capital portfolio risk and enterprise-wide risk (ERM) is a significant trend in the early years of the twenty-first century with British firms, such as Euro Log, remaining at the forefront of European public and private sector concepts and practice in supporting the culture, organisation, processes and procedures for managing project uncertainty.

These examples demonstrate the influence of contexts on the innovation process and the continuing emphasis of project managers on a particular set of tools and techniques to manage complexity and uncertainty.

Alongside these innovations in scheduling and risk management tools, project managers experimented with different organisational structures for achieving greater integration of functional specialisms, which were a feature of the classical approach. At one end of the organisational spectrum is the project as part of the functional divisions of the organisation. At the other end of the spectrum is the pure project organisation, where the project is separated from the rest of the parent system. The project team becomes a self-contained unit with its own administration, but tied to the parent company by varying degrees of overall strategic direction, administrative, financial, personnel, and control procedures, and reporting of progress.

In the 1960s, the matrix organisation was introduced in an attempt to encapsulate the advantages of the project organisation structure and functional organisation. Essentially, it comprised a pure project organisation overlaid on the functional divisions of the parent organisation. Knight (1977) identifies three forms of matrix structure:

- the coordination model or 'lightweight matrix'
- the overlay model and
- the secondment model or 'heavyweight matrix'.

The first of these is seen as being the weakest form of matrix structure as there is often limited commitment to project success by the members of the project team and the influence of the project manager can be less than the functional heads. The overlay model seeks to balance the influence of the project manager with that of the functional heads. In the secondment model, the functional departments second people on a full-time basis to the project team and on completion of the project they return to their line function.

Out of the need for greater integration came systems management practice whereby performance requirements were specified, the project carefully pre-planned to prevent future changes, and a prime contractor was appointed to be responsible for development and delivery. The concept of the systems support contractor emerged as an alternative to the traditional practices. It involved a new approach of specifying the total system and designing components to perform within it, allowing a complete weapon system to be planned, scheduled and controlled, from design through test, as an operating entity. This demonstrates the growing realisation of a more holistic approach to project management influenced by systems thinking. However, there were weaknesses in the approach with continuing emphasis on dividing the tasks as the primary responsibility for the project was passed to the contractor.

The great urgency of the Atlas missile programme to counter the threat posed by the perceived progress of the Soviet Union in developing long-range ballistic missiles in the 1950s led to the introduction of concurrent working in project management. The practice of testing major systems associated with the programme simultaneously rather than consecutively can be seen as an important factor in the introduction of concurrent working. The Atlas programme also demonstrated the importance of strong and effective leadership and championing in order to achieve project targets on time to meet the intense pressure to change associated with the Soviet military threat.

The Polaris programme initiated in the 1950s further increased the influence of the project group over the functional hierarchies associated with classical organisational structures. As in the case of the Atlas missile programme, it further demonstrated the need for a holistic approach – in this project seeing the weapon as a single system comprising the submarine and the missiles. In other words, defining the total project was seen as being increasingly critical to success. This total system approach also recognised the need to take into consideration the external environment, including forecasting and accommodating the rate of technological change in the external environment. This included addressing external factors such as sensitive political issues and building the broad base of political support necessary for the project. It illustrated the importance of taking into account the complex interaction between internal and external factors when introducing innovation. The way the project was set up and managed addressed the growing evidence that ineffective leadership and championing and a failure to anticipate and correctly estimate the change in the external environment were the causes of many project failures.

The importance of people and leadership issues in projects had been identified as long ago as 1959, by Gaddis who published a seminal article, 'The Project Manager', in the *Harvard Business Review*, as reported in Morris (1994). In this he identified a number of key issues in the management of projects such as the project manager's style, the need for organisational support, the need to take operational decisions for the benefit of the overall project, the importance of addressing conflict in projects, and the problems of authority and responsibility. The need for organisational integration, teamworking and commitment were also identified in the 1960s in the Apollo space project and its management systems. Morris (1994) argues that, to many, Apollo became the model of modern project management. The people involved employed a range of increasingly sophisticated techniques, concepts, approaches, philosophies and practices that were much more than the development of a set of management tools and techniques. The technological challenges were immense, with the very precise but challenging objective of transporting a man to the moon and returning him safely to earth by the end of the decade, and all within a budget of $20 billion. The technical, schedule and budgetary objectives were clear and broad project objectives were established within which detailed planning could take place. Given its high profile, the Apollo programme became very influential in introducing and promoting innovation in project management, and confirmed the importance of flexible responses to internal and external factors including political, economic, social, legal and technological.

During the 1960s, there was an increased interest in project management with the establishment of professional project management associations and increasing numbers of books and articles on the new discipline. This increasing status of the emerging discipline corresponded to the greater adoption of matrix organisations and the use of projects in high-tech, construction, development aid and other industries and in military and civilian sectors. This greater formalisation of the knowledge associated with project management can be seen as the recognition of the growing significance and complexity of projects and the importance of their effective management within an expanding number of industries and economic sectors. The concepts of modern project management began to be diffused into construction during this period but it was the techniques that tended to be adopted rather than its broader concepts and principles.

With the increasing emphasis on environmental issues and concerns over the future of the planet during the 1970s, an increasing number of projects were abandoned because project teams had largely ignored the external factor of environmentalist opposition. It became increasingly clear that the inward looking, organisational and procedures-driven approach to integration and systems management was no longer sufficient. As discussed in the next section on post-Fordism, there was a growing realisation that a much wider range of external factors needed to be embraced by project teams. This meant that economic, social, political, cultural and ecological factors needed to be more formally integrated into the project management process. Morris (1994) argues that this was particularly evident in transport, nuclear power, North Sea, Third World and space projects. He further argues that project management failed to recognise this shift and remained focused on clarifying middle management issues such as control systems, the power and authority of project managers, conflict and organisational structure. There was also a failure to recognise the poor success rate of many projects with cost and schedule overruns resulting mainly from poor forecasting in the feasibility stage. Attempts to introduce post-Fordist approaches to innovate and manage complexity were further impeded by the continuing, and indeed increased, emphasis on rigid and price-competitive approaches to managing projects and the opportunistic behaviour of the key participants.

The on-going Fordist orientation of project management is illustrated through an analysis of its main definitions. Most of the literature on project management describes projects as characterised by the following features.

- An objective which leads to the creation of some new entity which did not exist before – a project is usually a one-time activity, which is goal orientated with a well-defined set of desired end results; an output or product which is typically defined in terms of cost; and the quality and timing of the output from the project activities.
- Uniqueness – a project is usually a 'one-off' and not a repetitive undertaking. Even 'repeat projects' will have distinctive differences in terms of the requirements of the client or project sponsor and the circumstances under which it will be undertaken.
- Uncertainty – the uniqueness of projects leads to uncertainty.
- Complexity – projects bring together many different functions and sometimes organisations to undertake a wide range of tasks. This can lead to complex relationships between people, functions, tasks and organisations. An additional aspect of complexity is the demands that projects make on a range of resources, usually on an intermittent or varying basis and within a particular set of constraints.
- Interdependency – projects are made up of a large number of separate but interdependent tasks, which often interact with other projects and the organisation's ongoing operations.
- Temporary nature – projects have a defined beginning and end, so a temporary concentration of resources is needed to carry out the undertaking. Once the project is completed the resources are usually redeployed.

- Life cycle – the resources needed for a project change during the course of its life cycle. From planning and control perspectives it is therefore necessary to divide the life cycle into project phases. Projects have life cycles similar to organic entities. Generally, from a slow beginning they progress to a build-up of size, then peak, begin a decline and finally must be terminated, which suggests that like most organic entities they often resist termination. Some projects end by being phased into the normal ongoing operations of the parent organisation.
- Linear and sequential process – projects are often described in phases with clear boundaries that follow a linear progression.

In addition to these key features, Lock (1996) and Woodward (1997), also highlight the idea of novelty and innovation which they see as the principal identifying characteristics of a project. Projects are viewed as a step into the unknown, fraught with risk and uncertainty because no two projects are ever exactly alike. Even a repeated project will differ in one or more commercial, administrative or physical aspects from its predecessor and therefore require some degree of innovation and learning to manage the unknown and uncertainty. However, while the uniqueness of projects is recognised, Slack *et al.* (2000) argue that, to a greater or lesser extent, all projects have some elements and characteristics in common which can be exploited to improve performance through innovation.

Although these key features recognise the interdependent nature of project management there is a strong tendency to draw hard boundaries around projects and between different phases of projects, which often leads to fragmentation and conflict. There is also a strong emphasis on measurable outputs limited to time and cost while ignoring very important but less tangible factors. These key features suggest that project management throughout the 1970s and 80s was still significantly dominated by the principles of Fordist thinking. Even many of the formal definitions of projects in the 1990s were still strongly influenced by Fordism. For instance, the British Standard BS 6079:1996, *Guide to Project Management*, defined a project as

> *a unique set of coordinated activities, with definite starting and finishing points, undertaken by an individual or organisation to meet specific objectives within defined schedule, cost and performance parameters.*

It went on to define project management as

> *the planning, monitoring and control of all aspects of a project and the motivation of all those involved in it to achieve the project objectives on time and to the specified cost, quality and performance.*

The Chartered Institute of Building's Code of Practice for Project Management, first published in 1992, defines it as

> *the overall planning, coordination and control of a project from inception to completion aimed at meeting a Client's requirements in order to produce a functionally and financially viable project that will be completed on time within authorised cost and to the required quality standards.*

These definitions still describe project management in terms of a rigid and linear process comprising stages such as planning, monitoring, control, etc., with an emphasis on tools to manage the harder, more tangible aspects of projects. This is also clear in relation to definitions in the context of construction. Walker provided a detailed definition for construction projects in the late 1980s:

> *The planning, control and coordination of a project from conception to completion (including commissioning) on behalf of a client. It is concerned with the identification of the client's objectives in terms of utility, function, quality, time and cost, and the establishment of relationships between resources. The integration, monitoring and control of the contributors to the project and their output, and the evaluation and selection of alternatives in pursuit of the client's satisfaction with the project outcome are fundamental aspects of construction project management.*

Again, there is clear evidence of Fordist influences in this definition, as discussed earlier with a strong recognition of construction's predominantly one-off projects and short-term relationships. Most construction projects, by definition, imply the achievement of a unique change, which should meet the needs of a client and a group of project stakeholders. As was seen in Chapter 2, the traditional approach to procurement had fragmented the design and construction process. In addition, with the move to more outsourcing and subcontracting, increasingly construction projects began to involve staff from a larger number of diverse organisations. As mentioned in Chapters 1 and 2, it became common for as many as 100 independent organisations to be engaged in even a modest construction project. The relationship between the project and its participating organisations is shown in Fig. 5.1. However, Walker's definition shows some recognition of the growing pressure on construction organisations and project teams to focus more on the external customer – the client – and the need for a more inclusive approach and a blurring of the boundaries between key project actors and activities. Features that are increasingly being promoted as a means to address the fragmentation of construction since the early nineteenth century, and hence its continuing disappointing performance.

Figure 5.1, however, is a simplified view of projects in construction as it shows only one project being undertaken by the same organisations. In

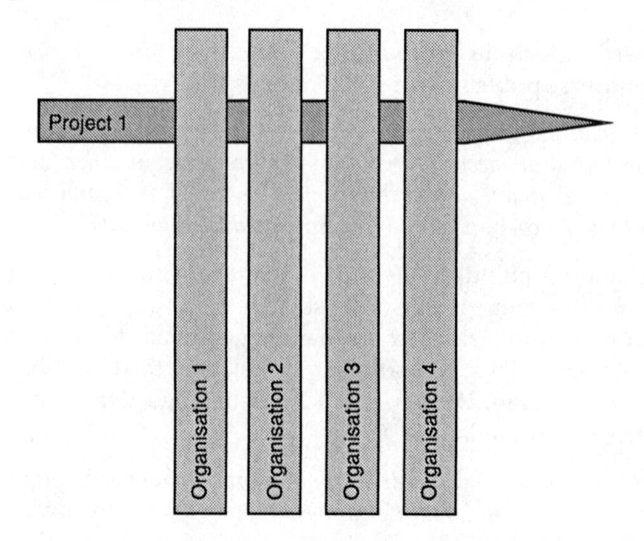

Fig. 5.1. Construction project involving a number of independent orgnisations

reality, as illustrated in Fig. 5.2, this may not be the case as there will often be a number of projects with different mixes of organisations involved in each project. The differences between the organisations involved are also exacerbated by the wide range of professional groupings and professionals involved. Figure 5.2 is also a simplification of the

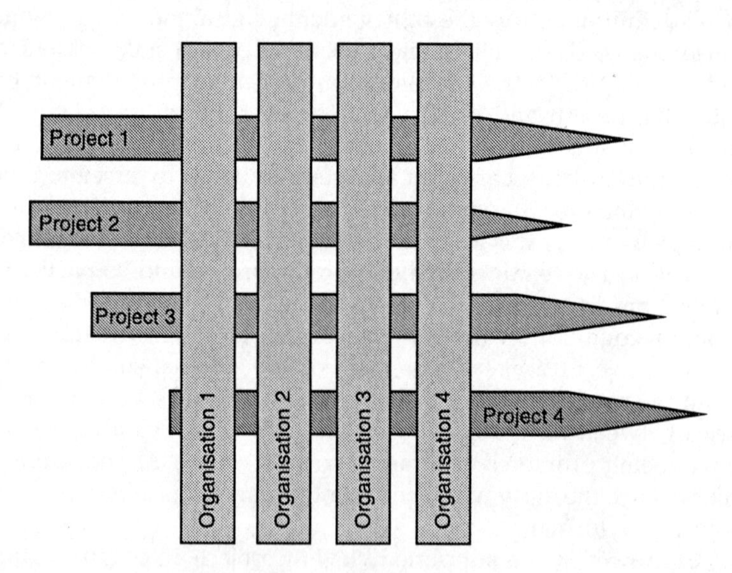

Fig. 5.2. Complex matrix of firms and projects

normal situation in that all the projects are shown as involving the same organisations. In reality, it is more far more usual for projects to be made up of people from a totally different mix of organisations. Each organisation will be driven by its own agenda and objectives and its behaviour by its culture. These differences create coordination and communication problems, often leading to conflict and adversarial relationships between the autonomous organisations within each project. The constant mixing of organisations across projects also presents a significant barrier to learning and the transfer of knowledge between projects and organisations. This complex matrix structure comprising organisations and projects raises two significant management issues:

- managing in the culture and context of the firm or organisation, and
- managing in the context and culture of the project.

The challenge for project management in this situation is to bring together individuals from quite different organisations and cultures and form them into an effective project team within the often relatively short durations of most construction projects. However, project management in general and construction in particular continues to be greatly influenced by Fordist thinking in this task. This means that much of contemporary project management has largely failed to develop effective teamworking contingent on the goals and strategies of the key stakeholders and participants. It has largely ignored the importance of individual and organisational behaviour and culture and, as a consequence, failed to provide a means of unifying overall processes and aligning individual, organisational and project objectives. It has also been largely unsuccessful in developing the synergy needed to support effective learning, knowledge sharing and innovation.

Matching project management with post-Fordist principles

As already discussed, the discipline of project management has attempted to bring together teams to solve the complex problems associated with the late twentieth century. Although its development straddles the transition from Fordism to post-Fordism, much of project management it is yet to fully embrace the main principles and practices of post-Fordism. This explains the growing body of literature and debate in relation to the effectiveness of project management in responding to the changes in the wider environment associated with the new paradigm. It also helps explain the obvious tensions in the project management community where some practitioners remain wedded to Fordist approaches and tools, with others recognising the need for the 'softer', more customer and people-oriented approaches associated with post-Fordism.

In the mid-1980s, there were substantial developments in our understanding of projects and ways to improve performance through innovation that began to match the emerging features of post-Fordism. In part, this was as a result of the recognition of the discipline and the influence of innovations taking place in key sectors of the economy such as the automotive and electronics industries. A more strategic and proactive role for project management also began to emerge to complement its previously reactive role mainly at operational level. A wider range of tools and techniques was developed to assist project managers to adopt a more holistic approach to the management of projects at strategic and project levels. Technology began to be used more effectively through, for example, prototyping and configuration or change management. Advances in computing hardware and software enabled more efficient cost, schedule planning, networking and other project management-based techniques to be used more easily and effectively. Simultaneous Engineering, Just In Time (JIT), Total Quality Management (TQM), Business Process Reengineering (BPR) and Lean Production emerged in a number of industries but particularly in manufacturing. As a consequence of the intra- and inter- organisational integration associated with these innovations, partnering and teamwork became increasingly important. A combination of the increasing complexity of many projects and the developments in the understanding of risk and its implications meant that risk management became a distinct and increasingly important discipline in the management of projects. With the increasing emphasis on continuous improvement and benchmarking associated with TQM and customer focus measures of project success become more common, even though the measures remained essentially to do with the harder aspects of projects such as time and cost.

The development of project management became increasingly influenced by systems thinking with its emphasis on understanding the 'whole picture' – including the internal and external environments of projects. As a result, projects are increasingly being seen as open systems interacting with their wider environment – including governments, competitors, suppliers and customers – nationally and, increasingly, globally. There is an increasing awareness that the external environment to projects is shaped by legal, economic, social, political, cultural and technological forces and that major projects can also have a significant impact on their external environment. This more holistic and open view of projects is increasingly challenging the more traditional approaches to project management. It also presents further challenges to project managers in that they now need to be much more aware of the interaction between economic, social, political and sustainability issues in the wider environment and the interrelationship between these issues and their projects. As well as being open and interacting with their external environment, projects are increasingly seen as embracing soft systems that need to recognise and can accommodate human aspects such as emotional reactions, personal values and shifting expectations. However, recognition of the systems approach remains largely focused more on the harder systems with insufficient recognition of the 'softer' and less tangible issues associated with most projects. It could be argued that hard systems being precise, well-defined and quantitative, are more readily accepted by project managers who continue to be more comfortable with situations where measurements are possible, numerical models can be constructed and outcomes predicted using tools with a degree of certainty and regularity.

During the early years of the post-Fordist era there was an increasing recognition of the importance of people in the success of projects. Carter *et al.* (1984), recognised the importance of motivating, involving and training people as prerequisites for initiating, implementing and sustaining change in projects. They related, for instance, a number of key elements to successful and innovative project management including:

- building the team and developing the interest of the people involved
- negotiating and agreeing common objectives
- addressing their practical, personal and social needs and
- training them to undertake the task and to envisage and initiate new uses of resources.

Although they also highlight the importance of monitoring, measuring, reviewing and evaluating performance, they failed to emphasise their links with continuous improvement. With the inclusion of supply and logistics in their model they also linked performance improvement and a removal of waste with a better integration of project participants and their supply systems.

Further changes needed in project management have also been identified by Pinto and Slevin (1987) and Slack *et al.* (2000) who highlight the significance of factors associated with the success of a project in the new paradigm such as learning, feedback capabilities, problem solving, customer focus and communication. These also include recognising the need for interpersonal as well as technical and administrative skills and the commitment of top management that has been effectively communicated to all parties. They also include performance improvement which is linked to feedback capabilities, problem solving and control mechanisms. In addition to the involvement of people identified by Carter *et. al.* (1984), Pinto and Slevin (1987) suggest the need for the involvement of key project personnel throughout the life of the project. This longer-term view of projects recognises the need for an increasing number of key project stakeholders to be involved for more of the life cycle – in some cases from conception through to obsolescence and recycling.

The substantial developments in project management thinking outlined above were increasingly supported through the setting up of a supporting infrastructure including organisations such as the International Project Management Association (IPMA), the Project Management Institute (PMI), and the Association of Project Managers (APM) in the UK – the latter becoming the Association for Project Management. Both PMI and APM have developed and are continuously refining bodies of knowledge, defining the areas in which a project manager and project teams should be competent. The discipline was further recognised through the certification of project managers. There is also a growth in the number of publications relating to project management and degree programmes containing more substantial elements of project management have been launched. During this period, project managers and their associations have begun to discuss 'management by projects' rather than the more restrictive term 'project management' (Morris, 1994).

Despite project management's developing support infrastructure and its long period of evolution, it can be argued that most of its concepts and techniques are still inadequate to the overall task of managing projects successfully in a more global and competitive post-Fordist environment. Many projects continue to be unsuccessful. The British Standards Institute (1996) acknowledges that the overall record of British organisations in managing projects leaves much to be desired. Maylor (2003) argues that this is not just a British problem and that 80% of projects worldwide are not delivering what was required of them in one or more substantive ways. This suggests that there are a number of significant aspects missing from our present understanding and practice of project management, and that the organisations involved are not deriving the full competitive advantage from their projects. There is still, for instance, a lack of awareness of the need for more synergistic and collaborative relationships between the project participants and more integrated and seamless project processes. There is also a need for better coordination, integration and synchronization between project teams and their supply systems.

The above analysis would suggest that project management needs to embrace further changes if it is to accommodate the shift to the challenges of the post-Fordist paradigm. In the new paradigm, project management is no longer about managing the sequence of steps required to complete the project on time and to budget but working concurrently on all aspects of the project in multifunctional teams, improving relationships and increasing integration between project teams and their supply systems in order to increase the focus on the needs and aspirations of both internal and external customers. This new approach to the management of projects involves establishing its objectives, defining the project strategy and setting it up to maximise the chances of it being successful for all its parties (or stakeholders), and accomplishing its goals efficiently and effectively. Morris (1994), argues that much of the past thinking on project management remains misfocused, and he suggests changing the title of the discipline from project management to the management of projects to reflect the shift to the new paradigm. He also argues for the continued removal of barriers around projects as the project's definition both affects and is affected by the external environment such as politics, community views, economic and geophysical conditions, availability of finance, and project phasing and duration.

The importance of people in the success of projects is also acknowledged by Morris who argues that projects will be much harder to manage and may be seriously prejudiced if the attitudes of the parties essential to their success are not positive and supportive. This more holistic and modern approach is clearly more ambitious and demanding and is seen as a new and powerful way to integrate organisational functions and motivate groups to achieve higher levels of performance and productivity. It involves the setting of appropriate objectives and much more subtle definitions of success. It requires management of some considerable sophistication in dealing with a broad range of specialised and strategic matters, ranging from finance, through wider environmental factors and technology to procurement and people management.

Other forms of project management models in the post-Fordist era are being developed which acknowledge the changing nature of projects and the interaction between the main elements of a project and its environment in which social factors such as leadership, communication, culture, learning, empowerment and motivation are given greater prominence. These models demonstrate the need for a robust approach to facilitate the integration of multiple project perspectives and levels of concern (Cicmil, 2000) and achieve greater alignment between the project goals and the multiple factors that influence project performance. Such models can also help to identify and address the instability and ambiguity in project environments and the explicit and often rigid nature of project constraints, particularly within one-off projects. However, these models need to acknowledge the principal factors that must be managed for projects to be successfully accomplished in the post-Fordist era. For example, the implications of these changes in the nature of projects and their management on the types of individual and collective learning needed in project management. This learning needs to be focused on unbounded systems thinking, revaluation of competitive advantage, risk assessment and management, and the evaluation of the areas critical to project success. This indicates a substantial broadening of the scope of project management as well as the need for considerable technical and social development.

Increasingly, the new form of project management is seen as a significant tool to managing change and innovation within and between organisations as they respond to the challenges of the new paradigm. This means that today's project managers, as well as having to cope with complex projects, are also being challenged to think of themselves as change agents – individuals who implement change by gaining the commitment and action of key people both within and between organisations. With the shift to post-Fordism, it could be argued that, given the flexibility of the emerging models of modern project management, it is well placed to identify and satisfy the diverging needs of customers and shifting environmental conditions in the new paradigm (Saad and Jones, 1995). In achieving this, however, project management needs to move away from simply considering time, cost and quality objectives to a broader, more stakeholder-oriented set of measures linked much more closely to business drivers and strategies with the early and synergistic active involvement of the key project participants and stakeholders. This new agenda indicates a shift within the McKinsey 7-S model (see Fig. 5.3) to the four soft Ss, staff, skills, style and stakeholders, rather than focusing primarily on the hard three hard Ss of systems, strategy and structure.

This means that, in spite of making significant steps towards post-Fordist thinking there remain a number of limitations to project management. This suggests a new agenda for project management development if it is to fully adopt post-Fordism and innovate to manage complexity

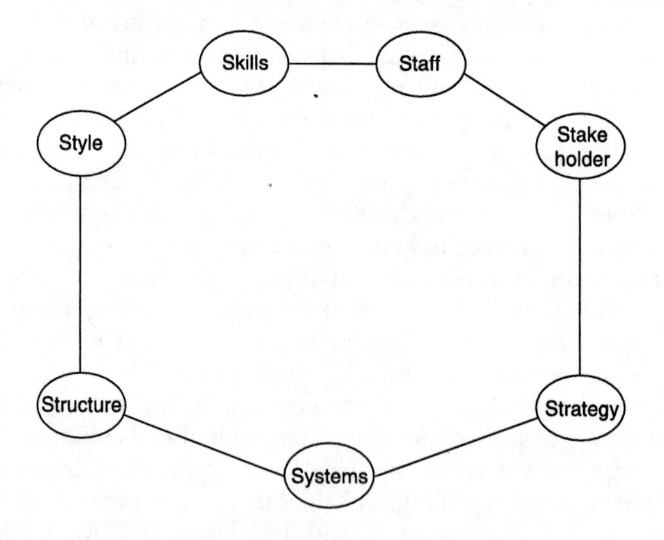

Fig. 5.3. The McKinsey model

and increase its competitiveness. For example, there needs to be a greater acknowledgement that project organisations are dynamic and open systems that interact with the external environment. The new agenda also requires that both internal and external customers, whose demands are having far-reaching consequences including the need for more complete solutions, are listened to and empowered. There is also need for a more inclusive approach involving more stakeholders and downstream participants through more synergistic relationships to improve performance and increase trust. In addition, there needs to be a recognition that successful implementation of this new agenda is strongly dependent on greater integration and openness between customers and suppliers, with greater transparency and more sharing of information and collective learning to continuously innovate and improve performance. Many aspects of this new agenda are currently being addressed through new approaches such as partnering and supply chain management, which are discussed in Chapters 6 and 7.

A review of the progress of project management in construction towards post-Fordism

The adoption of post-Fordism in the management of construction projects has followed, to some extent, a similar pattern to other sectors of the economy. There have been a number of significant innovations such as Fast Build, Construction Management and Build Operate Transfer (BOT) aimed at introducing alternatives to the sequential and fragmented thinking that continues to dominate much of construction since the introduction of the traditional approach to procurement at the beginning of the nineteenth century. Since the hyper-inflation of the 1970s, fast-track, a form of concurrent engineering requiring considerable integration and synchronisation across the project process and between firms is practised in many projects. From the late 1980s, more collaborative relationships have been gradually introduced through project-specific partnering. However, as outlined in Chapter 3, there are some specific features of construction projects and their management that are shaping and indeed impeding the way such post-Fordist approaches are being accommodated in the industry. As in other parts of the economy, project failures still persist because of the large number of autonomous organisations involved in the temporary coalitions associated with most construction projects. This results in continued reliance on fragmented processes and adversarial and contractual relationships. Project teams and processes are established in a number of ways but most are based on price-competitive procurement systems, which continue to dominate the industry. This means that a number of different potential processes and supply systems have to be provided but only one set will be selected to deliver the project's work packages. As a consequence, considerable effort, time and cost is spent in preparing and evaluating bids, the majority of which will be unsuccessful, rather than focusing on adding value for the end-user. Also, many of the project's key participants, such as main and specialist contractors and material and component manufacturers and suppliers are not involved in the critical decision-making in the early stages of projects.

Managing construction projects involves many organisations from a range of industrial sectors working together in a project team on project-specific tasks. The project-based nature of work, coupled with the limited role of main contractors in the traditional approach to procurement, means that they operate primarily as merchants purchasing, assembling and managing a number of complex supply chains from diverse supply systems and selling the inputs on to the client rather than developing a product for the client. The rigid division of organisations into project and corporate business groups as described earlier causes tensions as the people involved must address both project and business objectives, which are often highly differentiated. Gann and Slater (2000) argues that business processes are, in general, ongoing and repetitive whereas project teams and processes are essentially temporary and unique. Organisations can develop practices in their business that can stimulate innovation and provide opportunities for continuous and sustained improvement. On the other hand, project processes are usually non-routine and require collaboration between autonomous organisations assembled for that particular project aimed at specific user and client needs. This results in the formation of temporary and transient project teams, which do not lend themselves easily to systematic repetition, standardisation and continuous improvement of processes and products.

Although the transient and temporary nature of construction project coalitions can be seen as a weakness of the industry, this approach to organisation does mean that construction organisations have the potential opportunity to develop a degree of focus on the customer impossible in many other industries, such as those that supply consumer goods. Such temporary project structures present considerable opportunities for flexible and innovate responses to project goals as long as the appropriate project participants are brought together in a creative environment and are willing to accept the additional risk associated with innovation in the product and/or process. Unfortunately, these circumstances remain the exception in the majority of construction projects. The demand for construction organisation's products and services fluctuates widely in response to business and investment cycles. This means that relationships with clients remain essentially short term, contractual and opportunistic with limited scope for the development of mutual understanding, shared learning and innovation.

Gann and Slater (2000) argues that because construction firms have traditionally had short-term and, at best, opportunistic relationships there has been insufficient consideration of the links between the performance of firms and the performance of projects. As already mentioned, project management has tended to focus on improving project performance in terms of project design, planning and execution. Given its continuing reliance on traditional procurement approaches it is hardly surprising that less attention has been paid by project managers to understanding how organisations make their inputs into projects, how they assemble and coordinate their activities and resources at the project level, and how to measure their effectiveness. The continuing use of price-competitive tendering and arms-length relationships demonstrates that there is also still insufficient recognition that the effectiveness of projects is dependent on a wide range of organisational characteristics including the financial position, business goals and objectives, structure, ownership, people and, most importantly the culture of the organisations involved.

As seen in previous chapters, there are a number of interrelated external factors in the construction market and wider environments that are increasingly impacting on project management and driving innovation. The globalisation of markets and production associated with post-Fordism have increased pressure on construction to change. Increasing demands for more economically, environmentally and socially sustainable developments have also begun to shape the way in which projects are undertaken. The structure and timing of finance is a growing influence on the planning of projects and procurement approaches. New and more complex ways of financing projects and the role of financial institutions are becoming increasingly important drivers of change. The diffusion of information and communication technology, new methods of manufacturing, and growth of the service sector have all affected demand for new types of buildings and development processes. Regular private and public-sector clients have begun to demand that their construction projects are delivered with greater certainty in relation to time and budget and to a wider range of quality requirements more closely focused on supporting their core corporate objectives. They are also demanding increasing flexibility from the industry with tailoring of project cycles to more closely reflect their core business cycles and needs, including shorter lead-in times and more appropriate whole life cycle characteristics. As construction market segments become more differentiated, the responses of project managers need to be more focused and contingent. In some segments of the market the ability to assemble teams quickly and flexibly to deliver one-off projects remains important, whereas in others the building of stable supply chains in order to improve performance in product and process over longer periods of time is a priority. Project management needs to respond to these changing circumstances and support problem-solving and value-adding solutions across a widening range of client requirements.

Project managers are attempting to improve teamworking within projects in a number of ways in order to respond to the challenges outlined above. This includes the appropriate selection of more cooperative team members on the basis of a number of criteria, using workshops to develop closer relationships, and co-locating the main project participants. Where a high level of cooperation and integration is achieved, projects can become more like organisations with bonds of competence and even goodwill, trust and common governance. However, this is still rare in the price-competitive approach to procurement that characterises most approaches to projects in construction. Indeed, the poor performance of many construction projects suggests that conventional project management has not always been able to develop the level of teamwork needed to cope with the dynamics of innovation in project-based firms in the new paradigm – mainly because it has largely failed to integrate project and business processes and objectives effectively.

This implies that the knowledge and types of skills needed by project managers have not kept pace with the radical changes associated with the new paradigm. Interdisciplinary design knowledge and integration skills are required to enable the more synergistic coordination, integration and synchronisation of the wide range of stakeholders and specialists involved. New skills include:

- strategic planning
- technical skills
- market analysis
- deal-structuring
- design and engineering
- systems integration
- economic and business case assessments
- inter-organisational relationship building
- legal, financial and procurement advice.

This increasingly requires the harnessing of skills and knowledge from a number of disciplines including engineering, technology and the social sciences – the 'hard' and the 'soft' aspects of management.

Although most of construction has yet to address the requirements of the new paradigm, the changes in the construction market and wider environment have begun to influence the response from parts of the supply side of the industry as construction organisations seek competitive advantage in the new business environment. These leading organisations have begun to reshape the roles of project participants and accept greater responsibility for more significant parts of the project life cycle, including the financial aspects and dealing with political and regulatory issues, as in the case of the Private Finance Initiative and Prime Contracting. These market demands have resulted in a number of new approaches to procurement and the lengthening of the involvement of the supply side of the industry in more of the life cycle of projects. There has been a growth in the demand for single-point responsibility or turn-key solutions, particularly with the introduction and expansion of Design and Build and Prime Contracting. Through the building of new business relationships associated with partnering, PFI and Prime Contracting construction organisations are having to develop longer-term, closer, more collaborative and trusting relationships with other organisations in their supply systems. This indicates that, in some parts of the industry, the relationships associated with the traditional approach to procurement between organisations in construction projects are changing. There is a move away from traditional arms-length relationships, where the focus is on individual organisations with clear and rigid boundaries, towards more coordinated mechanisms between firms, including co-engineering practices and more collaborative relationships.

As a consequence of the need to develop greater customer focus, some construction organisations are concentrating their efforts on specific clients, types of client or market sectors. They are becoming more specialised as they build expertise in relation to specific markets such as social housing, retail, leisure, etc. This demonstrates that some leading construction organisations are responding positively and enthusiastically to a specific aspect of post-Fordism by developing their technical and management capabilities in a limited number of areas. For example, designers Arup utilise their reputation to deploy technical expertise as their competitive advantage, and the contractor Pearce Retail exploits its ability to work more collaboratively upstream with a rationalised client base and downstream with its key specialist and trade contractors, manufacturers and suppliers. In this way, construction organisations can be seen to be responding to changes in technology, the market and, more specifically, to the increasing complexity of client needs, and social, economic, technological and political circumstances.

However, project-based organisations face considerable challenges in introducing and sustaining the changes outlined above within the form of project management based on post-Fordism. Innovation is being impeded by the continuing use of price-competitive and contractual approaches to procurement by many clients and their advisers, and by construction organisations often putting themselves forward for inappropriate projects. Where performance improvements have taken place in projects during the 1990s, clients are much more thoughtful and careful in their approaches to procurement and supplier selection and suppliers have responded by aligning themselves more closely and synergistically with a smaller number of their key clients. Managing innovation in the majority of construction projects remains difficult because of the number of organisations involved, their short-term, contractual and opportunistic relationships and their complex organisational boundaries. Much of project management in construction remains concentrated on tools to manage cost and time and, hence, has largely ignored tools to manage relationships and innovation.

Knowledge in construction projects is differentiated but in the new paradigm it needs to be shared by project participants and supply networks to ensure successful innovation. The integrity of the information and knowledge and the management of know-how in these networks of suppliers is an increasingly important strategic consideration for both customers and suppliers as they innovate to seek competitive advantage. Yet project management has provided construction organisations with insufficient time and little incentive or indeed the capabilities to share information within such temporary coalitions and to address the problem of organisations using knowledge opportunistically by retaining it within their own sphere of control.

Another fundamental difficulty in innovating in construction is the discontinuities of project-based production, which results in broken learning and feedback loops on performance. Prototyping and experimentation are also difficult to justify in the context of one-off projects, which means that computer-based simulation and modelling is of great importance at the front end of construction projects. Indeed, such developments in IT have allowed more virtual simulation, allowing clients to better understand proposed buildings, although this has yet to be adopted in the majority of projects. ICT has also provided a tool to address the uncertainty of projects and the interdependency between different parts of the building and its participants. It has helped design and construction teams in product definition, development, simulation, testing and production where extensive amounts of technical information and knowledge have to be transferred between the large numbers of specialists involved. Although there has been this development of IT, the fragmented nature of project teams and their supply systems and their lack of social cohesion is still impeding the shared learning and innovation needed for competitive advantage in the new paradigm.

Although the informal and largely unrecorded learning, research and development often undertaken within construction's more complex and unique projects should not be underestimated, the continuing strong influence of Fordism makes these difficult because of the often considerable barriers to the transfer of knowledge needed for innovation from organisations to projects and between organisations. The management of innovation in construction is often conducted informally with few mechanisms for identifying and capturing progress. Such management that exists tends to be confined to R&D units, senior management and engineering staff within construction firms and, much more commonly, within supply systems. However, the new paradigm is associated with a more holistic, inclusive and integrated approach, requiring the continuous generation and sharing of knowledge. Learning needs to take place within both the organisations and the projects. Organisations need to identify the core competencies necessary to support and enhance the knowledge and skills of their project teams. Business processes tend to be ongoing and repetitive while project processes are unique and short term. Hence, project managers need to identify, capture and integrate these different forms of experiences in order to better achieve the project's objectives and satisfy the needs of the customers. This alignment of organisational and project learning is difficult in the context of the short-term relationships associated with conventional project management in much of construction.

Although some leading clients and construction organisations are increasingly encouraging learning and innovation within projects and seeking to use it to gain competitive advantage, unfortunately, many people within the organisations involved are often unaware of this activity or actively discourage it because of the rigid contractual arrangements and the opportunistic relationships they have with project participants. It is hardly surprising, therefore, that construction should now be making such significant efforts to improve the links between project-based firms and their supply networks and between project-based firms and their customers. These links are seen as being increasingly important in ensuring that the more sophisticated products and services demanded by clients can be designed, integrated and operated. More innovations flowing up from the specialist and trade subcontractors and their supply systems are also being encouraged as more and more responsibility for design is migrated downstream. In order for this to work there must be more involvement and empowerment of key participants downstream in the process (Jones and Saad, 1998), better flows of information and knowledge relating to the clients' requirements down the supply chain, and better flow of products, services and value up the chain to the client. This is placing more emphasis on the need for greater standardisation of components and processes offering the possibility of meeting differentiated and customised client requirements while benefiting from the economies of scale derived from the use of standardised pre-assembled components.

Construction is also developing much-needed measures of project and corporate performance in order to drive innovation and continuously improve performance. However, these have essentially remained focused on the more tangible factors relating to time and cost. Organisations need to find more appropriate mechanisms to interpret and evaluate the key project-based activities. However, this is beginning to be addressed with the introduction in 1988 of the national set of key performance indicators (KPIs). This first set of overall KPIs was a landmark for the industry. Their development was a recognition that if the industry as a whole is to make the quantum leap in performance improvement which is needed to become more competitive then it has to measure its performance and do so on a common basis. The purpose of the KPIs is to benchmark a project or company against the range of performance currently being achieved across the industry. They provide a means for clients to ask potential suppliers to provide information about how they perform and assess their suitability for a specific project. Construction supply chain companies can benchmark their performance to enable them to identify their strengths and weaknesses. The KPIs allow the government to measure the industry's improvement over time. There are ten headline KPIs, seven of which refer to project performance, and three are measures of company performance.

(*a*) Project performance
- client satisfaction – product
- client satisfaction – service
- defects
- predictability – cost
- predictability – time
- construction cost
- construction time.

(*b*) Company performance
- profitability
- productivity
- safety.

Although this is a landmark development in construction, it is still very focused on the measurement of tangible factors as it continues to place more emphasis on cost and time and ignores intangible factors that can promote and sustain motivation, learning and innovation. Although tools are frequently used to track time and cost in projects, they are hardly used to track value-adding innovations. Innovations are rarely examined, costed and evaluated in order to help provide strategic direction for the organisations involved or future projects.

In order to address less tangible aspects of performance improvements, some construction organisations are attempting to adopt a more holistic approach including 'soft' and 'hard' issues through the use of the EFQM (European Foundation for Quality Management) model for Business Excellence. While the approach recognises the importance of key business results such as profit, return on capital employed, shareholder earnings and achieving budgets, it also introduces the importance of leadership, motivation and internal and external customer satisfaction. It is based on the principle that, in order for an organisation or team to succeed, there are a number of key enablers on which it should concentrate to achieve improvement goals. It also needs to measure its performance through a number of key results areas. The key enablers include how well the organisation is led, how well its people are managed, how far its policy and strategy are developed and implemented by its leaders and people, and how well it manages its resources and suppliers, and develops and manages its processes. The key results areas by which the EFQM model then measures successes are how far it satisfies its customers, how well-motivated and committed is its workforce and how the local and national community outside the organisation views its activities in terms of its contribution to society.

Conclusion

Projects are important for individuals, organisations and economies and there have been many successful projects where effective project management has delivered better customer relations and control, with shorter development times, lower costs, and higher quality, reliability and profit margins. Project management can provide a sharper focus on results, better functional integration and higher morale. On the other hand, many projects fail with a significant body of evidence to show that there are problems with projects and their management as most projects are delivered late or over budget or both. Project management can increase complexity, increase cost, result in more conflict and lower motivation of project participants. Also, there are a number of critical issues to be addressed by project management in responding to the new post-Fordist paradigm and the key features of innovation as identified in Chapter 4. Increasingly, there are also questions as to construction's ability to successfully implement and sustain innovation, given the short-term and adversarial relationships, and the tensions and trade-offs between time, cost and performance that characterise its current approach to managing projects. Innovation aimed at total customer satisfaction may not be as straightforward or realistic a proposition in the context of construction projects. Their uniqueness, uncertainty and risk and the difficulties of planning for time, budget, material and human resources – given that the idea and expectations for a one-off project are only best estimates of what the future might hold – can be seen as significant inhibitors of innovation.

As shown in this and previous chapters, innovation in construction is inhibited by a number of critical factors which include:

- poor understanding of the project purpose, its business case and end-clients/user needs
- limited capability to manage the uncertainty of the external environment
- the lack of top management commitment
- poor management of risk
- lack of communication of the project's overall goal, objectives and strategy
- opportunistic and often adversarial relationships
- limited review of project performance and transfer of learning between projects
- fragmentation of processes
- different organisational cultures
- dynamics of organisational environments
- barriers to collective learning and knowledge transfer and
- limitations of resources at any point in time.

The current mind set in much of project management in construction suggests that it cannot fully respond to the changing view of construction projects. Conventional methods developed over the last 60 years or so are too cumbersome and lack sufficient focus on customers and the people and organisations involved. Many project managers are failing to recognise that projects are no longer seen merely as a complex sequence of activities independent of the wider environment which progress through stages of the project life cycle and unfolds in a linear-rational manner to deliver clearly defined objectives using well-defined methods. Increasingly, they involve the consideration of environmental issues and the synergistic integration of all participants and processes. Projects include addressing client/user/stakeholder concerns and aspirations and setting up relationships and modes of effective communication within the project team, between the team members and their organisations, and between project teams and the wider community.

There is a growing demand for more valuable bundles of services and products, which is blurring the boundaries between the different project participants and key project stages. Implementing and sustaining innovation to deliver these new products and services requires new management approaches and business practices requiring a radical change in attitudes of all those involved leading towards greater collaboration, learning and flexibility which conventional project management has largely failed to support in most construction projects. It also needs a greater focus on developing collaborative customer-supplier relationships, less adversarial and fewer contractual arrangements to encourage inter-organisational networking, sharing and learning. Chapter 6 examines how partnering is emerging as an alternative strategy for innovation in some parts of construction as it brings stakeholders together in more stable and synergistic ways to improve project performance, achieve total customer satisfaction and build greater mutual competitive advantage.

References

BADEN HELLARD, R. *Project Partnering: Principle and Practice.* Thomas Telford, London, 1995.

BRITISH STANDARDS INSTITUTION. *Guide to Project Management.* BS6079, BSI, London, 1996.

CARTER, R., MARTIN, J., MAYBLIN, B. and MUNDY, M. *Systems, management and change.* Paul Chapman Publishing in association with The Open University, London, 1984.

CICMIL, S. Quality in project environments: a non-conventional agenda. *International Journal of Quality and Reliability.* LOVE, P. E. D. and GUNASEKARAN, A. (eds). **17**, No. 4/5, 2000, 554–570. Bradford: MCB University Press.

GADDIS, P. O. The project manager. *Harvard Business Review,* May–June 1959, 89–97.

GANN, D. and SLATER, A. Innovation in project based service-enhanced firms: The construction of complex products and systems. *Research Policy* 2000, **29**, 955–972.

GULICK, L. Notes on the theory of organisation. In *Papers on the Science of Administration.* URWICK, L. (ed.). Institute of Public Administration, Columbia University Press, New York, 1937.

JONES, M. and SAAD, M. *Unlocking Specialist Potential: a More Participative Role for Specialist Contractors.* Thomas Telford, London, 1998.

KNIGHT, K. (ed.). *Matrix Management – A Cross-Functional Approach to Organisations.* Gower Publishing, Aldershot, 1977.

LOCK, D. *Project Management.* 6th edn. Gower Publishing, Aldershot, 1996.

MAYLOR, H. *Project Management.* 3rd edn. Prentice Hall, Edinburgh, 2003.

MEREDITH, J. R. and MANTEL. *Project Management: A Managerial Approach.* 3rd edn. John Wiley & Sons, Chichester, 1996.

MORRIS, P. W. G. *The Management of Projects.* Thomas Telford, London, 1994.

PINTO, J. K. and SLEVIN, D. P. Critical success factors in successful project implementation. *IEEE Transactions on Engineering Management,* **34**, No. 1, 1987.

SAAD, M. and JONES, M. *New organisational and cultural arrangements in the management of projects in construction in vision to reality.* Australian Institute of Project Management, National Conference, 1995.

SLACK, N. *et al. Operations Management.* 3rd edn. Pitman Publishing, London, 2000.

WALKER, A. *Project Management in Construction.* Blackwell, Oxford, 1989.

WINCH, G. M. *Managing Construction Projects.* Blackwell Science, Oxford, 2002.

WOODWARD, J. F. *Construction Project Management.* Thomas Telford, London, 1997.

Chapter 6

The emergence and implementation of partnering in construction

The adoption of partnering can be seen as construction's first significant attempt to innovate through addressing the industry's adversarial relationships and fragmented processes. This involves developing more stable and long-term relationships with a smaller number of preferred suppliers, which, in turn, provides greater continuity of work for their suppliers of construction products and services. Other key objectives in pursuing this approach includes addressing opportunistic behaviour, power imbalances and lack of equity, high levels of defects, cost overruns, and contractual and confrontational procurement strategies leading to disputes. This chapter assesses the motives, relevance and the effectiveness of partnering with reference to the key features of successful implementation and management of innovation as highlighted in Chapter 4.

The theoretical origins of partnering

The concept of partnering emerged essentially in the management of operations in the automotive industry. Following Toyota's lead, for some time major car assemblers had been developing longer-term, closer and more collaborative relationships with their key suppliers. They had embarked on this in order to decrease inventories (these include raw materials, parts and components and semi-finished products in the production process, and finished goods ready for shipment), improve quality and decrease the costs of production by integrating processes and breaking down the barriers between the firms involved.

Partnering is also derived from the partnership philosophy which emerged with globalisation. The world is increasingly being characterised by a trend towards greater complexity, uncertainty and diversity. At the same time economic activity, business organisations and technology are changing fast, resources are getting scarcer and markets are being increasingly deregulated. As a result, organisations are looking outside their internal environment for ways of dealing with these changes. Different forms of collaborative alliances are used to respond to these changes and to achieve a 'critical mass' of financial and human resources, greater growth and competitive advantage, and greater flexibility and responsiveness to new opportunities. The main purpose is to enhance collaborative advantage and to produce long-term mutual benefits for each partner (Lamming, 1993 and Mowery 1988). This inter-organisational alliance aims to have long-term mutual benefits for each contributing partner. It is formed by organisations with other organisations, which can include suppliers, customers and, occasionally, competitors and even higher education and research institutions (Dodgson, 1996). It requires sharing risks, responsibilities, resources and rewards, all of which can increase the potential of collaboration beyond other ways of working together.

The agreement of a common purpose and a clear definition of objectives constitute the most fundamental elements of partnering. Improved mutual understanding, which is both an important prerequisite and benefit of partnering, is based on open communication and sharing of ideas, concerns and information. It also motivates individuals, groups as well as organisations in their search for continuous improvement and the achievement of their objectives. It can prevent conflicts and foster a cooperation bond and a spirit that will ensure that individuals, groups and organisations understand each other and are driven by common objectives that are mutually beneficial (Bennet and Jayes, 1998).

Definition of partnering in construction

Unlike partnering in other sectors, the main objective in construction is to improve performance through building better coordination and longer-term relationships leading to greater trust. The Construction Task Force report (Egan, 1998, p. 12) *Rethinking Construction* defines partnering as follows:

> *Partnering involves two or more organisations working together to improve performance through agreeing mutual objectives, deriving a way of resolving any disputes and committing themselves to continuous improvement, measuring progress and sharing the gains.*

The Construction Industry Institute (CII) developed an early definition of partnering as:

> *a long-term commitment between two or more organisations for the purpose of achieving specific business objectives by maximising the effectiveness of each*

participant's resources. The relationship is based on trust, dedication to common goals, and an understanding of each other's individual expectations and values.

A later definition, by the Reading Construction Forum (RCF):

Partnering is a managerial approach used by two or more organisations to achieve specific business objectives by maximising the effectiveness of each participant's resources. The approach is based on mutual objectives, an agreed method of problem resolution and an active search for continuous measurable improvements.

The Construction Industry Board (Working Group 12) has provided a more recent definition of partnering. This builds on the Latham report and the work of the Reading Construction Forum. It states that

Partnering is a structured management approach to facilitate teamworking across contractual boundaries. It should not be confused with other good project management practice, or with long-standing relationships, negotiated contracts, or preferred supplier arrangements, all of which lack the structure and objective measures that must support a partnering relationship.

The CII's definition emphasised 'long-term commitment' and 'trust' – two factors which are significantly lacking in traditional contracting strategies. This definition implies a fundamental and radical shift in the traditional business relationship between clients and contractors. Although the definition given by the Reading Construction Forum recognises the importance of 'mutual objectives' and 'an agreed method of problem resolution', it has omitted 'long-term commitment' and the role of trust. This definition can be seen as proposing a more realistic and pragmatic approach to partnering which acknowledges that the majority of construction's clients are infrequent users of its products and services. It could also be seen as providing a more realistic view of the confrontational relationships between clients and contractors that have been established and reinforced over many years through the use of the traditional approach to procurement. The third definition, given by the Construction Industry Board, attempted to reduce the level of confusion and misunderstanding associated with the concept of partnering by articulating the key elements of the concept and differentiating it from other approaches.

These definitions show that the motives for adopting partnering in construction are different from those in manufacturing. Whereas the emphasis in construction was on ending disputes and improving relationships and trust, in other sectors the main focus was on reducing waste and adding more value. The definitions also show that the perception of this innovation has evolved since its introduction into the UK in the late 1980s. This evolution can be explained by the industry's search for the most appropriate form of partnering contingent on given clients and their project participants and processes. This implies that all partnering approaches are to some extent different in terms of, for example, scope, aims, objectives, duration, value systems, content, focus, adoption, implementation and outcomes. This lack of a single definition or model of partnering can explain the misunderstanding of the concept and the misuse of the term.

Partnering in construction is a confused concept meaning different things to different people. To some it means a very close, single-sourced relationship, while to others it means effective project management. To some it has been a great success while for others it has been a complete failure. This suggests that partnering is not a unified concept and that it takes a number of different forms. Also, that it might not be appropriate for all situations. It would also indicate that, like most innovations, successfully implementing and sustaining it is very difficult. Table 6.1 illustrates some of the forms of partnering that have been identified in construction in the UK. This helps to explain the number of definitions of partnering and the considerable confusion in relation to what it really means and entails.

In spite of the diversity of views of partnering, there are underlying principles which are common to all 'true' partnering approaches. These include a commitment to promote more positive and collaborative relationships and a common purpose leading to mutual advantage. The main overarching aim of partnering is to increase the certainty that the customer's needs are met, and that the suppliers that have contributed to the process feel adequately rewarded for their efforts. The main factors differentiating it from the project management

Table 6.1. Forms of partnering in construction

Forms of partnering	Relationship duration	Basis of partner selection
Project-specific	One-off, project by project	Competition with some negotiation
Post-award project-specific	One-off, project by project	Competition
Coordination agreement	One-off, long term	Competition, negotiation
Preselection arrangement	One-off, long term	Negotiation
Strategic or full	Long-term	Competition and negotiation

Table 6.2. The greater emphasis on intangible and soft issues

Project management	Partnering
Procedures, systems, tools, techniques, schedules	People, relationships, common purpose, trust, commitment
Tangible and hard issues	Intangible and soft issues
Shift →	

normally associated with construction contains the inclusion of more organised and formal efforts to improve relationships by reducing conflict and the need to resort to litigation. Other factors include improving communication, increasing quality, adding value, and increasing profitability. Another major difference is the greater emphasis on building more openness, transparency and trust in relationships – between individuals and firms – than would normally be associated with traditionally managed construction projects. It is also important to appreciate that partnering is not just good project management. It needs to be seen as going beyond the normal project management associated with construction projects and introducing a new culture based on improved relationships and trust leading to a win–win situation. The aim of partnering is to provide project management with a more appropriate framework aimed at managing the soft issues in order to enhance performance and competitiveness through better relationships, trust, openness and empowerment.

The nature of relationships in construction and partnering

As discussed in the previous chapters, performance and innovation in construction are significantly impeded by the adversarial relationships and fragmented processes that continue to dominate much of construction. Figure 6.1 shows the wide range of relationships and contracts available to clients and their advisers with the introduction of strategic partnering.

The vertical axis shows the range of possible relationships from arm's length through a number of increasingly closer and collaborative relationships to a strategic alliance (the closest relationship possible short of a merger of the organisations involved or a friendly takeover). The horizontal axis shows the range of contractual relationships from a

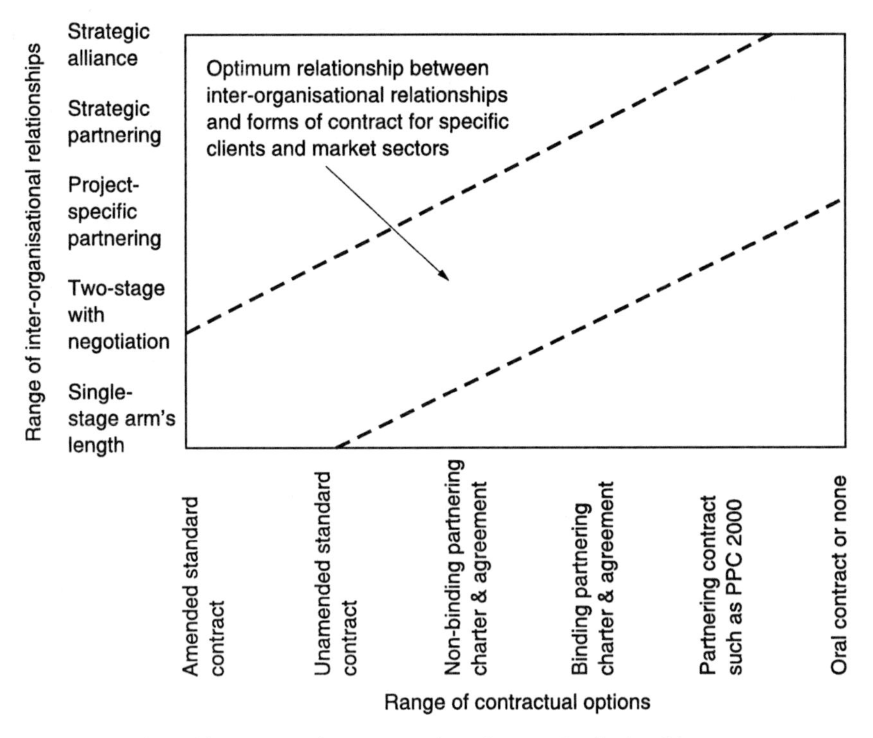

Fig. 6.1. The wider range of contractual options and relationships

firm standard contract, amended (almost invariably) to benefit the client and to shift risk to the contractor, through to a memorandum of understanding or an oral contract. The objective is to match the form of contract to the closeness and openness of the developing business relationship and the degree of competence trust which exists. Where considerable confidence in the business relationship exists and where there is a high level of competence trust between the parties there will be less need for a firm contract. Using this model, an optimal contracting strategy for a particular client's partnering approach can be planned and implemented in an open and transparent way that suits all the parties and over an appropriate period of time.

Figure 6.2 shows the relationship that most commonly exists between a client and their main contractors. It shows that, in this case, the client has arm's-length relationships with their main contractors and uses a standard form of contract, but which is made even tougher on the contractor by the addition of amendments. These amendments are drafted, by the client's legal advisers, to the benefit of the client by, for example, transferring responsibility for all risks to the contractor. As discussed in

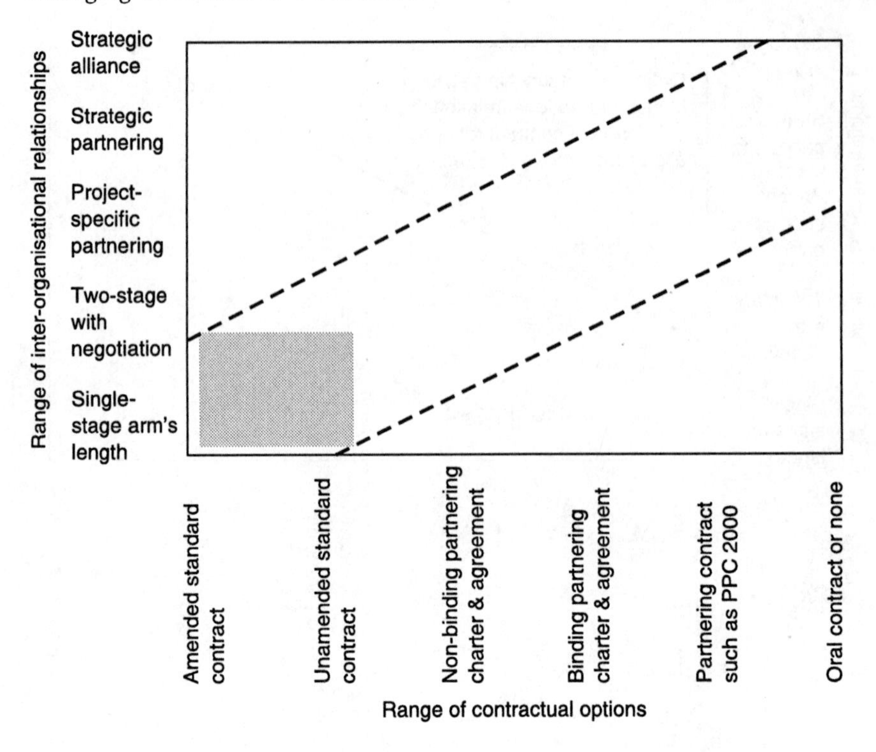

Fig. 6.2. A common contracting strategy

Chapters 2 and 3, the relationships between clients and main contractors show that the majority of construction's clients and their advisers favour such contracting strategies. Hence, most of the contracting strategies adopted by clients and their advisers can be plotted towards the bottom left-hand corner of the diagram, as shown in Fig. 6.2.

The standard forms of building contracts are not an ideal basis for forging more collaborative relationships between clients and their construction suppliers as they are closely associated with the adversarial relationships and lack of trust that have dogged the construction industry over many years. However, the role of the contract in building closer relationships is often overstated in construction for obvious reasons. A common way of setting up a partnering arrangement is to combine one of the standard forms of contract with a Partnering Charter and Framework Agreement – a method recommended in HM Treasury Guidance on Procurement. The Charter and Agreement can set out the true understanding between the parties in relation to the key aspects of their business relationship. As the relationship develops, the partners may wish to move towards one of the more 'intelligent' and collaborative contracts. These new contracts, such as PPC 2000, have been drafted to be more conducive to partnering than the more traditional forms of contract. As such, they are much less likely to be used as weapons at times when the relationship comes under strain. Although these new contracts seek more intelligently to underpin closer, more open and collaborative relationships associated with partnering there is still considerable nervousness about using them at this early stage in their development and use and in view of the industry's long history of opportunistic behaviour.

However, not all construction's clients adopt such potentially adversarial and short-term approaches to procuring their buildings and facilities suggested in Fig. 6.2. For example, Fig. 6.3 shows the very different contracting strategy adopted by McDonald's, the fast-food chain.

McDonald's have developed strategic alliances with two contractors for the development and construction of modular buildings. In addition, they have a single-source supplier for statutory approvals and development advice, and a series of preferred suppliers for the detailed design of restaurant components and site infrastructure work. McDonald's has no formal conventional construction contracts governing their relationships with module suppliers or its construction firms. They merely issue a purchase order incorporating its standard terms and conditions.

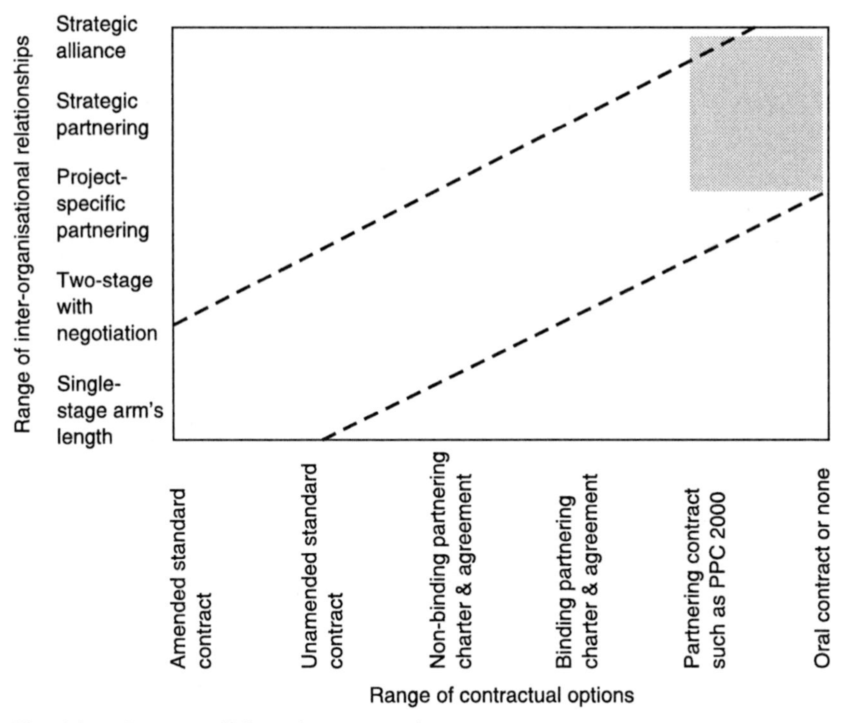

Fig. 6.3. A more collaborative contracting strategy

Case study 6.1: McDonald's and George and Harding – demonstration that relationships in construction do not have to be contractual

The £620 000 McDonald's restaurant in West Bromwich opened in December 2000, just 11 weeks after planning approval was gained, epitomising the product-team integration promoted by *Rethinking Construction*. Main contractor George and Harding (G&H) designed and constructed the project. They worked with Yorkon who supplied and assembled the factory-built modular building. The former filling station site required extensive stabilisation measures and suffered from unforeseen ground obstructions and atrocious weather conditions. Yet the project was completed within budget and on time and without a written contract.

McDonalds's senior project manager Gareth Jones explains how:

> *Working with a contractor should be like working with another department of McDonald's, where no contracts are required. Everyone works together in a team, each contributing to ensure the success of the project.*

G&H director Carl McCrudden adds,

> *Working with McDonald's and using M4I principles has given us a massive lead in understanding partnering. It's a lot more difficult than many people realise – we have no excuses. It does however create an extremely innovative working environment which allows ideas to flow freely from all parties without worrying about contractual relationships and it's much more fun.*

Partnering without a written contract potentially leaves G&H and McDonald's out on a limb legally, with only common law to fall back on should their informal dispute resolution process fail. But the parties see this as a small commercial risk. G&H chairman Colin Harding endorses partnering without a written contract. He argues:

> *As soon as one partner feels the need for some form of written contract, real partnering has ceased.*

The main benefits include the following.

- Capital cost – savings of £126 000 were made through innovative design and construction, £30 000 replicable in future McDonald's.
- Construction time – despite the site conditions being more difficult than expected the restaurant actually opened two

days early. Value engineering, quick assessments of options and rapid decisions were essential in keeping to budget and programme.
- Predictability – McDonald's projects run efficiently. The supply chain is geared to just-in-time delivery.
- Accidents – respect for people is an essential ingredient to make this tight logistical operation a success. There were no accidents at all in the 7000 hours worked on this site.
- Defects – G&H managers aim for and usually achieve zero defects at handover. Even with the accelerated delivery of the building, they cleared 90% of snags in one day and 100% within four days.

Movement for Innovation, Case History, www.m4i.org.uk

As already established in Chapters 2 and 3, this type of approach is still not widely used in construction where relationships remain mainly adversarial. There are a number of reasons which explain the strong and continuing preference for arm's-length, one-off, and contractually confrontational relationships fifteen years after the emergence of partnering. One of the main reasons is that the majority of construction's clients only build infrequently, which means it is difficult if not impossible for them to develop closer, more open and longer-term relationships with their construction suppliers. Another factor is the limited information relating to the actual cost of construction. Through a combination of a lack of such robust cost information and the lack of trust and openness in their relationships even many regular clients and their advisers believe price-competitive tendering to be the only safe approach to procurement. They argue that the only way the client can obtain a building or facility at its true cost is to take it to the market and put it through a highly price-competitive tendering process with as many as seven contractors competing aggressively for the work.

Unfortunately, this price-competitive approach adopted by the client and the advisers leads to opportunistic and adversarial behaviour by the successful contractor who seeks to obtain a reasonable profit from the project. As a consequence of the contractor's opportunistic and claims-oriented behaviour, in turn, the client and the advisers are forced into the position of exercising even greater surveillance of the contractor, and when it comes to the client's next project it is hardly surprising when he and his advisers decide to use a different contractor in an even more rigorous price-competitive tendering process. As a consequence, the client is more than likely to instruct the advisers to draft even tougher contracts and exert even more surveillance over the successful contractor in order to minimise the effects of their *anticipated* opportunistic behaviour.

The direct outcomes of repeating this cycle over a number of projects include the industry's contractually confrontational and adversarial culture and its unenviable reputation for the number of disputes leading to litigation – all of which leads to a further erosion of trust between clients and their advisers, and contractors. However, the situation is even worse than this gloomy analysis suggests. In persisting with this approach, the client and the advisers and the main contractor effectively forego the opportunity for continuous improvement through innovation in the delivery of construction projects in terms of cost and time reduction, and enhancements in quality and safety.

The emergence of partnering in construction

During the late 1980s and 1990s, more and more leading clients began to take a fundamentally different direction in their approach to procurement, mainly in an attempt to address the adversarial nature of relationships with their suppliers of construction products and services. This is shown in Fig. 6.4. This was led by the offshore oil and gas sectors and some major supermarket chains who had also begun to develop more collaborative but essentially informal relationships with their key suppliers. In adopting this strategy they were also following the example set by clients in the North American and Australian construction industries.

Slowly and carefully, these leading clients began to develop longer-term, more collaborative and less contractually confrontational relation-

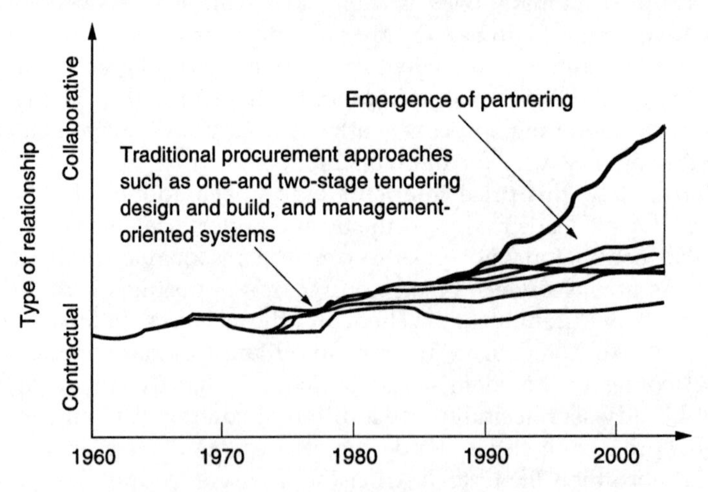

Fig. 6.4. The emergence of partnering in construction

ships with a smaller number of 'preferred' consultants and contractors under the banner of partnering. This constituted a radical shift from the essentially confrontational approaches that had dominated construction for many decades to 'win–win' solutions for the whole construction team.

Partnering was also given considerable additional credence and impetus by Sir Michael Latham's report, *Constructing the Team*, which was published in 1994. This greater emphasis on collaboration and integration was also strongly endorsed four years later, in 1998, by Sir John Egan's Construction Task Force with the publication of its report, *Rethinking Construction*, which identified 'partnering in the supply chain' as one of its four priorities for change.

Less confrontational contracts aimed at avoiding disputes, delays and ultimately extra costs – most notably the *Engineering and Construction Contract* (previously known as the *New Engineering Contract*) and PPC 2000 – have also been introduced. Each of these new approaches to procurement has, to some extent, increased the possibilities for greater integration and collaboration. However, despite these changes, the majority of relationships between clients and contractors have remained essentially short-term and contractual with some notable exceptions such as Marks and Spencer and Bovis who have had a more collaborative arrangement which has been successful and which endured over many years.

The different forms of partnering

There have been a number of attempts to define partnering in construction. A common way is in terms of its stages of evolution such as 'first' and 'second generation partnering'.[1] Another approach is to classify it in terms of the length of the relationship between the parties involved – such as project-specific and strategic.

'First generation' partnering began to emerge in the late 1980s in a number of sectors of the industry, but mainly in the retail and offshore oil and gas industries. Studies of these early ideas on partnering suggest that it revolved around three key principles, as shown in Fig. 6.5, which were applied within individual projects on a project-by-project basis (Bennett and Jayes, 1998).

More specifically, these three key principles involved:

- agreeing *mutual objectives* to take into account the interests of all the firms involved

[1] More recently, third generation partnering has also been identified, the key features of which can be closely associated with the features of supply chain management. This will be addressed in Chapter 7.

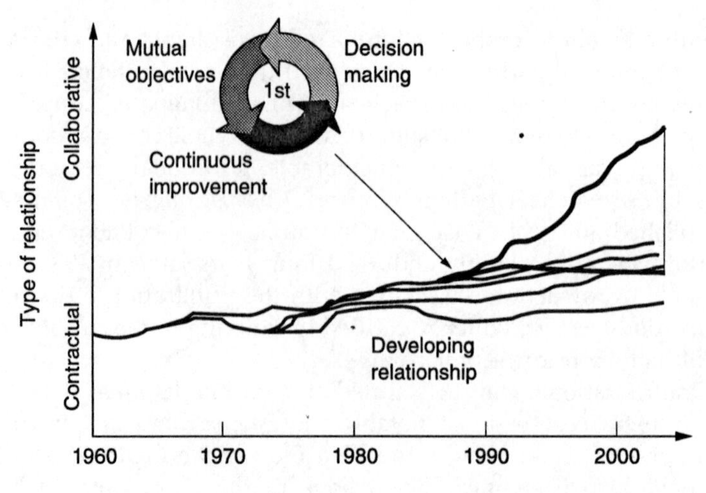

Fig. 6.5. First generation partnering

- improving communication and *making decisions openly* and *resolving problems* in a way that was jointly agreed at the start of the project
- aiming at targets that provided *measurable improvements in performance* from previous projects.

This form of partnering recognises the difficulties of building long-term relationships between clients and main contractors throughout much of construction given its traditional price-competitive, contractually confrontational, one-off approach to procurement. It also acknowledges that the majority of construction's clients do not commission buildings on a regular basis and therefore find it difficult, if not impossible, to exert enough leverage over their construction suppliers to persuade them to improve performance through longer-term relationships. This form of partnering is also perceived to be particularly well suited to public sector procurement, with its many checks, constraints and audit requirements such as the need to demonstrate probity, value for money and public accountability. Although this basic form of partnering has been shown to provide some benefits, it has limited scope for developing the relationships and integrated processes that are necessary for the substantial step improvements in performance that have been demonstrated in other industries.

Encouraged by their early success with first generation partnering, a number of leading-edge private-sector clients, with the advantage of substantial annual construction programmes, have progressed to 'second generation partnering'. A key element in making this shift was the client's ability and willingness to extend the relationship with fewer contractors beyond the duration of a specific project to a period of time or number of projects (known as a term or framework agreement respectively). These developments of both the scope of partnering, and the duration of the relationship between the parties are illustrated in Fig. 6.6.

The progression towards the adoption of the principles associated with greater collaboration is more evident with the emergence in the late 1990s of the 'second generation' style of partnering which includes a strategic decision to cooperate by the key project partners (Bennett and Jayes, 1998) as shown in Fig. 6.7. This more developed form of partnering often places greater emphasis on a more holistic approach based on a wider range of performance criteria in addition to time, quality and cost and acknowledges the strategic importance of such longer-term business relationships. It also incorporates some of the key features of the 'fifth generation innovation' such as developing stronger external linkages and inter-organisational relationships throughout the design and construction process. It is increasingly being seen as a way to develop a culture based on greater cooperation in longer-term and more stable relationships (Barlow *et al.*, 1997; Bresnen and Marshall, 2000), and as a way of addressing the industry's fragmentation and lack of integration (Bennett and Jayes, 1998).

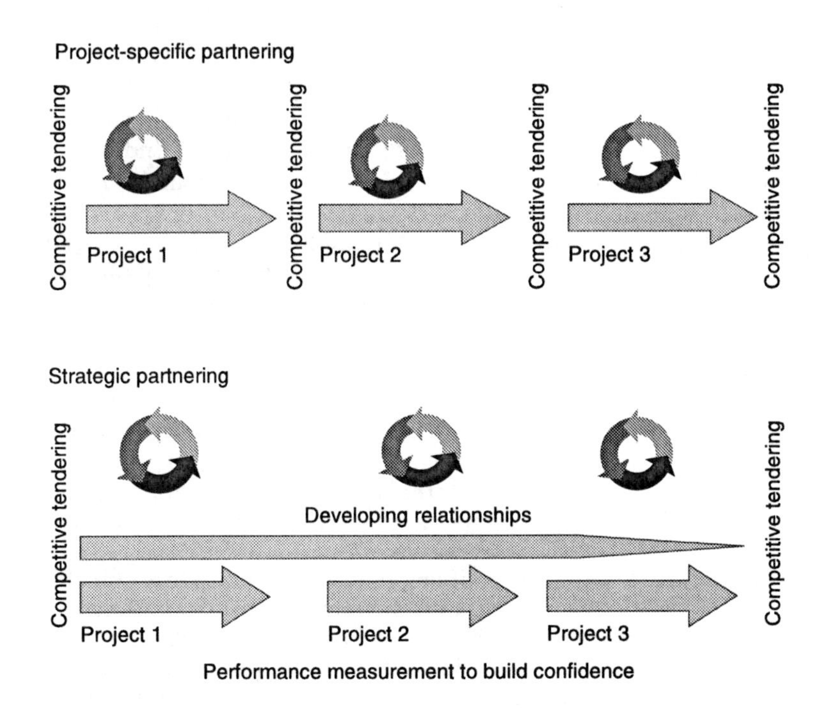

Fig. 6.6. Keeping the relationship going through a framework or term agreement

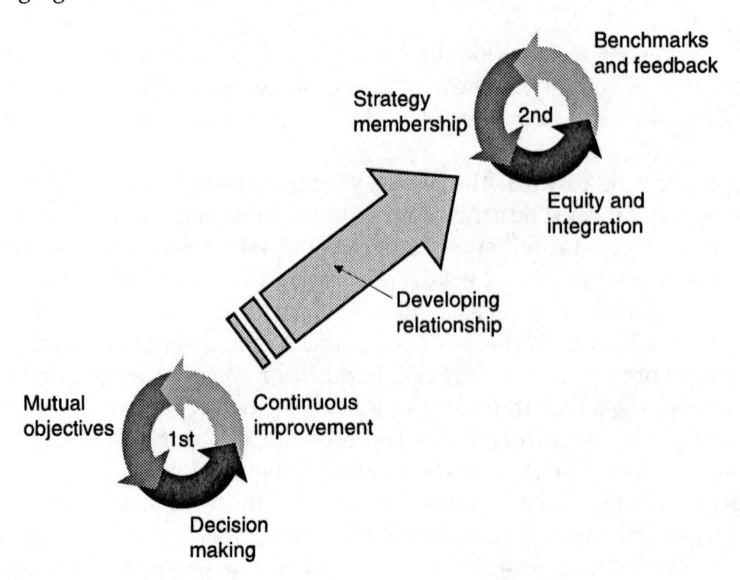

Fig. 6.7. Progression from first to second generation partnering

This 'second generation' partnering provides the opportunity to address the limitations of project-specific partnering by building stronger relationships and trust in order to integrate processes over a series of projects or a given period of time rather than within the constraints and limitations of individual projects. This more advanced and ambitious form of partnering comprises a number of features that can be associated with a fifth generation innovation as explained in Chapter 4. This drive to achieving greater coordination and integration has led to a shift from three to eight principles (Bennett and Jayes, 1998) as shown below. This suggests that this more advanced form of partnering is a much more complex and multi-factor innovation.

Strategy:	developing the client's objectives and how consultants, contractors and specialists can meet them on the basis of feedback
Leadership:	leading the overall partnering approach but also each of the business units within the relationship
Membership:	identifying the firms and business units that need to be involved to ensure all necessary skills are developed and available
Mutual advantage:	ensuring everyone is rewarded for their work on the basis of fair prices and fair profits

Relationship building:	improving the way the firms involved work together by cooperating and building trust
Benchmarks:	setting measured targets that lead to continuous improvement in performance from project to project
Process integration:	establishing standards and procedures that embody best practice based on process engineering
Feedback:	capturing lessons from projects and task forces to guide the development of strategy and performance improvement.

This can be seen as a significant step in developing the philosophy of collaboration to improve relationships, build trust and integrate processes through strategic partnering based on a win–win situation, strong leadership, performance measurement and continuous improvement. However, this second generation partnering has a number of weaknesses, including an undue emphasis on upstream relationships and a lack of involvement of key participants such as specialist and trade contractors and suppliers downstream in the process. This implies a continuing lack of empowerment and involvement of downstream participants.

It is important to be aware of the strategic and operational implications of the implementation of this second generation of partnering. The shift from 'first' to 'second generation' partnering must not be underestimated, and partnering on this scale should not be seen as just another procurement route but a fundamentally different way of managing and 'doing business' in construction. For instance, it requires a radical change in the relationship between the client and main contractor. On the demand side, it requires that clients and their advisers move away from contractual, price-competitive approaches to procurement with a large pool of competing contractors. On the supply side, main contractors and other suppliers need to respond by being more open and transparent in relation to business transactions, including divulging the actual cost of construction, and their levels of profit and overheads. In this way, clients and their advisers can continue to feel comfortable about not awarding work on the basis of competitive, fixed-price tendering. Contractors and other suppliers can also benefit from greater continuity of work and more certain and greater profits.

Robust measures of performance are needed if all parties are to feel comfortable with the new relationship and working together in this more collaborative way. Measurement and benchmarking are also needed to counteract supplier complacency, demonstrate continuous improvement and continuing value for money, and justify higher levels of profit and other rewards for suppliers. It requires the careful identification, allocation and management of the risks that, if not properly managed, could lead to major conflicts and the breakdown of the relationship. It is important that the volume and value of the business between clients and their suppliers does not lead to a dangerous mutual overdependency in the relationship.

In spite of the difficulties in implementing strategic partnering, an increasing number of construction organisations are attracted to strategic partnering, because of the growing evidence of both the tangible and intangible benefits from its successful implementation.

The benefits of partnering

Most recent evaluations of partnering show that improved construction performance in some cases greatly improved performance – is achievable. The Reading Construction Forum has estimated that project-specific partnering can deliver cost savings of 2–10%. Through their study of 200 examples of partnering, they found that the most sophisticated 'second generation partnering' approaches produced time savings of around 50% and cost savings of over 30%. But there are also less easily measured and intangible benefits for successful partnerships. These 'second generation' partnering relationships have provided the participants with:

- greater focus on the client and end-user
- improved flexibility to meet the changing needs of clients and end-users
- faster project start-ups
- better design, greater functionality and added value
- improved quality
- the reduction of defects at handover and during the design and construction process
- more stable workloads
- more predictable and higher margins
- lower accident rates
- more innovation.

Some 'third generation' partnering relationships, where the participants have undergone fundamental changes to attitudes and behaviour, organisation and technology, demonstrate even greater benefits.[2] Underlying these performance gains are the beneficial effects of softer and intangible improvements in communication and responsiveness to

[2] Third generation partnering will be examined in Chapter 7.

perceived problems, the development of greater inter-organisational and inter-personal trust, more awareness of opportunities to develop mutual advantage, and the promotion of a culture to support and sustain innovation and learning. The transfer of knowledge and understanding between the organisations involved in more advanced and collaborative relationships is also conducive to more effective shared learning and innovation. It provides the parties with an environment that allows them to refine and develop new competencies in a more controlled way with lower risk. A continuity of personnel from project to project in long-term partnering relationships also provides the opportunity to capture and transfer learning and innovation from project to project.

The following two case studies illustrate the mutual advantage to be derived from successful partnering. They outline how two early adopters of partnering were drawn to it and the benefits it has given them. The first provides a client's view – that of the retailer Sainsbury's. The second outlines partnering from a contractor's perspective – that of Pearce Retail.

Case study 6.2: Sainsbury's – a client's view of partnering

In the early 1990s, Sainsbury's faced external and internal difficulties to which they responded by adopting a process called 'Genesis' to generate greater success in the company. It created new challenges for its Property Services Division which had its staff reduced from 240 to 65 while its workload remained the same.

Outside Sainsbury's, things were also changing. Construction prices had been kept down by virtue of the depressed market and the strength of Sainsbury's position as a major client of the industry, but some of their many contractors were giving a poor performance while others were performing very much better.

Through a combination of a reduction in the number of its contractors, partnering, TQM, new and less adversarial forms of contract, and the setting of measurable performance targets linked to rewards, there was a 30% reduction in time and cost between 1991 and 1998.

When interviewed in 1998, Charles Johnston, Property Services Manager for Sainsbury's said, 'This approach has been a great success with huge potential for the future'. Further actions include encouraging cross-fertilization of ideas between projects and the use of continuous improvement groups to encourage learning and identify ways of doing things better.

Case study 6.3: Pearce Retail – a contractor's view of partnering

Like many contractors in the early 1990s, Pearce Construction was facing severe price competition resulting in uncertainty of workload and low profitability. Having analysed a number of studies in construction and approaches being adopted in other sectors of the economy, they decided to respond to the challenges by concentrating their efforts on three specialist areas of the construction market – including retailing – rather than chase after competitively tendered work across the whole sector. In this way Pearce Retail was formed.

In 1996, Pearce Retail began partnering with the supermarket chain ASDA, which had already begun the process of reducing and consolidating its main contractor base from 39 in 1994 to 4 in 1998. ASDA was looking for performance improvements from a much smaller pool of preferred contractors but across a broadening range of performance criteria including quality, flexibility, reduced impact of construction operations on the customer, faster times from inception to handover – and all at reduced cost.

The benefits of the partnership to Pearce Retail are that it now has more repeat work, the overhead associated with winning work has been dramatically reduced, profitability is improving, and its people have acquired new specialist knowledge and skills in a very significant sector of the construction market.

When interviewed in 1998, Andrew Staniforth from Pearce Retail said,

> *Working closely with ASDA has increased our confidence in our future workload. This is providing us with the scope to plan the growth of our company and implement essential changes in culture and processes aimed at adding greater value for our client and constantly enhancing our reputation.*

Both case studies demonstrate that partnering is still evolving, there is no single, or best way of partnering, and that organisations do it for different reasons. They show that, in specific circumstances, there are significant benefits for clients and suppliers from a successful implementation of partnering. However, as has already been stated, it is not realistic to consider partnering as being appropriate in all situations and under all circumstances. Also, a number of observers of inter-organisational relationships argue that actually all buyer and supplier objectives in critical areas will almost always be in conflict. Therefore, before embarking on partnering it is important that the possible pitfalls, problems and concerns are recognised and addressed.

Concerns associated with partnering

There are both fundamental weaknesses in the concept of partnering (Christopher and Jüttner, 2000), and substantial difficulties in defining (Cox and Townsend, 1998) and implementing it in the context and culture of construction given its transient relationships and overoptimistic views of human nature and motivation (Barlow *et al.*, 1997; Uher, 2000). Another major weakness is that most partnering in construction is currently predominantly initiated upstream by regular and experienced clients (Bresnen and Marshall, 2000), with less attention being paid to the role of specialist and trade contractors and suppliers downstream in the process (Uher, 2000; Barlow *et al.*, 1997). This suggests that the positive role that other participants such as specialist and trade subcontractors can make to projects has not yet been fully explored (Jones and Saad, 1998).

It is only more recently that leading private clients have begun to broaden their concept of partnering and to extend it downstream to include other key participants through supply chain management. Also, some public sector clients, have been seeking to emulate leading private clients by adopting Prime Contracting which was piloted in two projects for the Defence Estates, an agency of the UK's Ministry of Defence, during the late 1990s. It aims to replace short-term, contractually driven, project-by-project adversarial relationships with long-term, multiple project relationships based on trust and cooperation. This can be seen as a very ambitious step-change towards a collaborative approach as it includes experiments in reshaping project and supply chain structures relationships and processes. It is based on identifying one or more strategic supply partners in each key supply area, and working with them closely to improve the value that they add to the end product, and then delegating to supply partners the 'first tier' – the management of their own suppliers – the 'second tier'.[3]

In addition, as shown in Chapter 3, there remains a strong reliance on complex and rigid standard forms of contracts. These forms, with their provisions for damages or penalties, retention of money and withholding of payment, are seen as being the least conducive to collaboration (Cox and Thompson, 1998). In turn, the standard forms of subcontract used to appoint specialist and trade subcontractors downstream in the process also reflect the terms of the associated main contract. In many cases, the client (to the disadvantage of the main contractor) and the main contractor (to the disadvantage of the specialist and trade subcontractor) often

[3] For more information about Prime Contracting see the *Prime Contractor Handbook of Supply Chain Management* by Holti *et al.* (1999).

amend these standard forms rendering them even less conducive to downstream collaboration.

The Davis Langdon Everest survey discussed in Chapter 3 also shows that the two forms of main contract which are seen as being the most conducive to collaboration, the *New Engineering Contract (NEC)* and *GC/Works/1*, with their provisions for some sharing of rewards and joint working, were only used for 0.45% and 0.1% of projects by value respectively. This suggests that the vast majority of construction's clients still consider it necessary to resort to formal main and subcontracts to control their main contractors and specialist and trade subcontractors downstream in the process.

A major and fundamental concern is related to the fact that the majority of construction's clients are irregular commissioners of buildings and facilities. This implies major difficulties in providing sufficient quantity and continuity of work that would make it worthwhile for their suppliers of construction products and services to enter into closer longer-term relationships and to fundamentally change their ways of working. As a result, this type of client will find it difficult, if not impossible, to progress beyond 'first generation' partnering.

There is also the question of equitable relationships. Experience suggests that it is extremely unlikely that there can ever be absolute equality in any supply relationship – especially in construction with its deeply embedded hierarchical and class-based structure. Most trading relationships, whether they incorporate partnering approaches or not, are driven by the relative status and power of the parties involved. In addition, there is a lack of understanding of the nature of power as either an enabler or inhibitor of partnering.

Another major concern is related to the need for a coincidence of interest between buyer and supplier, and the acceptance of the need for mutual competitive advantage. In a partnering arrangement, both parties need to feel confident of the other's openness, transparency, goodwill, integrity and commitment to helping each other to achieve project and business goals. This is perceived as being difficult to achieve given construction's continuing emphasis on price competition, its opportunistic and confrontational behaviour and deeply rooted blame culture.

There are also the dual dangers of complacency and over-dependency. As partnering develops there is a natural tendency for suppliers to become complacent in relation to full participation and continuous learning and improvement. The danger for suppliers is that they become over-dependent on a single client. The changing situation of Marks and Spencer provides a highly visible example of the disadvantages of suppliers making their assets specific to a single client or customer – however successful they are now or have been in the past.

increased vulnerability of the parties and their possible exposure to complacency and over-dependency.

(*d*) Developing the right organisational structure helps facilitate communication and interaction between individuals, functions and organisations. Experience shows that the quality of relationships, like the degree of closeness between the parties, is strongly influenced by the interface structure which has been set up to manage them on a day-to-day basis. Therefore, defining a balanced relationship and the development of the right interface structure are mutually reinforcing. Indeed, there appears to be no clear-cut attribution of effects between these two fundamental requisites.

(*e*) The next important prerequisite recognises the importance of motivating and empowering people throughout the whole process of implementation. Many companies struggle to implement partnering because of the resistance of people to change or their inability to perform their new role as a result of a lack of empowerment. Progress in developing inter-organisational relationships through partnering is strongly dependent on the relationships between individuals. Evidence suggests that emphasising internal relationships with employees and reinforcement of their commitment are strategic prerequisites to closer external relationships. Gaining such employee commitment, however, is a difficult and complex task.

(*f*) As well as intra-organisational learning, inter-organisational learning is also necessary if the full benefits of closer and more collaborative relationships are to be fully exploited. However, the mere existence of a partnering relationship is not in itself sufficient to stimulate and support shared learning and the transfer of knowledge between the parties. The degree to which this will occur is influenced by several factors including:

- the type of partnering being used
- the commitment of each party
- the nature of the knowledge to be transferred
- the openness of intra- and inter-organisational communication
- the culture of the organisations involved and
- their receptiveness to the new knowledge.

(*g*) The knowledge, skills and systems for monitoring the effectiveness of relationships are also important prerequisites. Closer relationships lead to greater interdependence between the parties and this, in turn, requires the development of appropriate metrics to monitor and measure the effectiveness of the more intangible and softer aspects of the relationship. Relationship monitoring refers to all procedures employed to evaluate whether the relationship meets the objectives of the parties. It involves joint negotiation and performance control activities and can be formal or informal depending on the specific circumstances. Independent monitoring may also be deemed necessary in the early stages of the development of the relationship and in those sectors of the industry where issues of probity and best value are very important – such as the public sector. The Reading University Estates Department's case study demonstrates the need for partnering and its successful implementation when the key prerequisites are in place.

Case study 6.4: Reading University Estates Department – an approach to partnering

After years of accepting lowest tenders, and achieving mixed results and some significant failures, the university decided there had to be a better way. They wanted more control over costs and better management of the inevitable design changes that arise as end-users (who are not professional construction clients) see the project develop. They ran Windsor Hall as a trial for procurement by partnering, which was relatively new in the university sector at that time, to determine whether partnering could produce a better value result.

The trial was a gamble with professional reputations at stake and the pressure to succeed was great. Fortunately, Mansell already had practical experience of partnering with BAA, and senior managers in Estates Services were impressed by their partnerships in the private sector and keen to learn how to emulate them.

The usual problems – information delays, late changes, shortages in the supply chain and sometimes complacent administration – did not go away. But how the team solved them made all the difference. Their approach included:

- professionally facilitated partnering workshops, team-building activities and a project charter (a commitment to 'good behaviour'), and regular partnership reviews to keep the team focused on core objectives
- committed leadership by principals in the client's agent and contractor groups
- project reviews organised so that problems were aired and resolved
- implications of changes examined in detail by the partners so that knock-on complications were foreseen and there were no cost and delay 'surprises'
- value-engineering workshops run package-by-package with end-users and contractors.

The simplest yet perhaps most effective change was talking about problems instead of writing letters, faxes and e mails.

The process of setting up the contract was novel at the time in university construction circles but is rapidly becoming best practice for occasional clients. Estates Services invited four reputable contractors to offer percentages for profit, overheads and preliminaries in the context of an estimated contract value and concept design. Mansell was narrowly ranked first on this basis. Contractors were also asked about their understanding and experience of partnering. Mansell won the contract following presentations to a panel of the client, end-users and client's agent. Personal judgement about which individuals 'one could confidently work with' was a significant factor in the final analysis.

The detailed design was developed once Mansell joined the team. Their paid input into the buildability of the design and value engineering led to 6% coming off the estimated cost. With 90% of the packages negotiated, the contract sum was struck at £3 086 000 and Mansell's percentage was converted to a lump sum. The team saved another 3% by further value engineering as the construction progressed.

The university awarded two further contracts to Mansell and tested other procurement methods along the way. They used the multi-party PPC 2000 for the Students Union (contracting with Mansell, the designers and key subcontractors) and a design and build contract with Mansell for the Archaeology building. They concluded that strategic partnering was the best solution for projects more than £500 000 in value and set up three framework agreements with Mansell, Interserve and Warings. Under these five-year contracts, which started in January 2002, each contractor can expect roughly one project a year. The university has contracted out its design services because it was no longer viable to retain in-house design specialists. Projects are increasingly judged on whole-life cost, not just capital cost.

Movement for Innovation, Case History, www.m4i.org.uk

Third generation partnering

The limitations and prerequisites identified in second generation partnering are to some extent being addressed by the emergence of third generation partnering, as shown in Fig. 6.8.

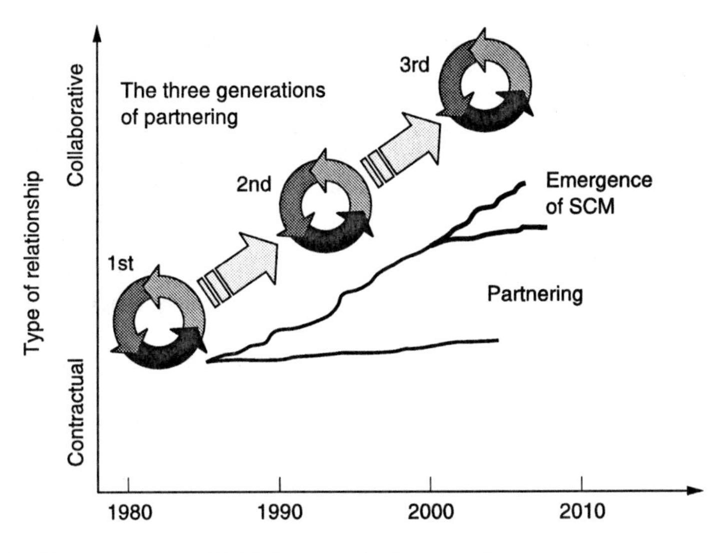

Fig. 6.8. The emergence of 'third generation' partnering

This more inclusive and challenging form of partnering corresponds to many features of supply chain management (SCM) which is investigated in Chapter 7.

References

BARLOW, J. COHEN, M., JASHAPRA, M. and SIMPSON, Y. *Towards Positive Partnering*. The Policy Press, Bristol, 1997.

BENNET, J. and JAYES, S. *The Seven Pillars of Partnering: A guide to Second Generation Partnering*. Reading: Reading Construction Forum, Reading, 1998.

BRESNEN, M. and MARSHALL, N. Partnering in construction: a critical review of issues, problems and dilemmas. *Construction Management and Economics*, **18**, 2000, 229–237.

CHRISTOPHER, M. and JÜTTNER, U. Developing strategic partnerships in the supply chain: a practioner perspective. *European Journal of Purchasing and Supply Management*, **6**, 2000, 117–127.

COX, A. and TOWNSEND, M. *Strategic Procurement in Construction: Towards Better Practice in the Management of Construction Supply Chains*. Thomas Telford, London, 1998.

COX, A. and THOMPSON, I. *Contracting for Business Success*. Thomas Telford, London, 1998.

DODGSON, M. Technological Collaboration. *Handbook of Industrial Innovation*. DODGSON, M. and ROTHWELL, R. (eds), Edward Elgar, 1996.

EGAN, SIR J. *Rethinking Construction*. Construction Task Force Report. DETR, London, 1998, p. 12.

HOLTI, R., NICOLINI, D. and SMALLEY, M. *Prime Contractor Handbook of Supply Chain Management Sections 1 and 2*. Tavistock Institute, London, 1999.

JONES, M. and SAAD, M. *Unlocking Specialist Potential: A More Participative Role for Specialist Contractors*. Thomas Telford, London, 1998.

LAMMING, R. *Beyond Partnership: Strategies for Innovation and Lean Supply*. Prentice Hall, New York, 1993.

LATHAM, SIR M. (1994) *Constructing the Team*. Final Report of the Government/ Industry Review of Procurement and Contractual Arrangements in the UK Construction Industry. HMSO, London, 1994.

MOWERY, DAVID C. (ed.). *International Collaborative Ventures in US Manufacturing*. Ballinger, Cambridge, MA, 1988.

UHER, T. E. Partnering Performance in Australia. *Journal of Construction Procurement*, **5**, No. 2, 2000, 163–176.

Chapter 7

The emergence and implementation of supply chain management in construction

Encouraged by substantial improvements in performance from first and second generation partnering, as discussed in Chapter 6, and the adoption of supply chain management (SCM) by other sectors of the economy, parts of construction have started moving towards the adoption of third generation partnering as shown in Fig. 7.1. This third generation partnering can be seen as corresponding in many respects to SCM and the key features of innovation needed for its successful implementation. The objective of this chapter is to assess the implementation and management of innovation in construction through the adoption of SCM.

The adoption of SCM is essentially being led by more informed private-sector clients who adopted first generation partnering during the late 1980s and early 1990s, and then second generation partnering in the late 1990s. As explained in Chapter 6, during the 1990s considerable effort went into defining and developing partnering to meet the different needs of the clients in various sectors of the construction industry. Unlike partnering, SCM is still very much perceived as a concept 'borrowed' from other sectors and which is in a very early stage of development in construction. It is also important to recognise that SCM is a complex innovation, which has proved to be difficult to implement and sustain even in sectors such as the automotive industry in which it originated. It is described (Saad *et al.*, 2002) as a fifth generation innovation which is:

- a multi-factor process
- built around close and long-term intra- and inter-organisational relationships
- requiring a strategic and long-term approach
- dependent on links with and support from the external environment
- necessitating continuous learning and
- commitment from top management.

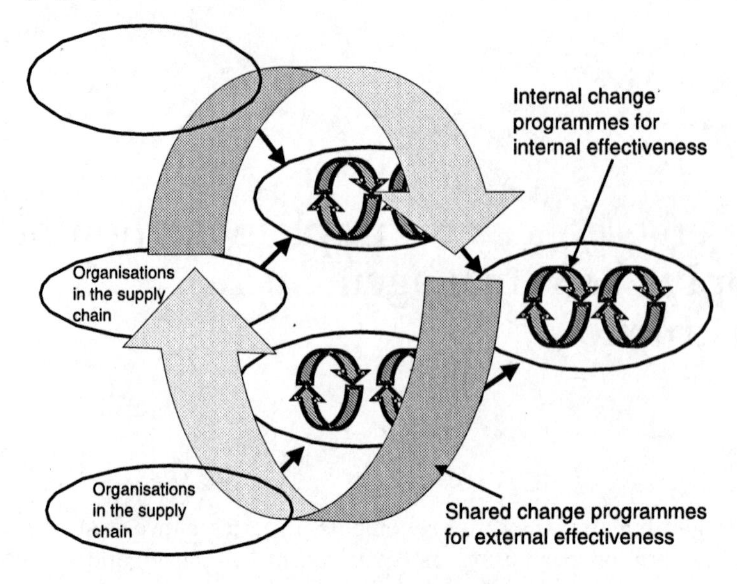

Fig. 7.1. Progression from internal to external programmes

Given the complexity of this type of innovation, this chapter reviews the concept of SCM before addressing its relevance to construction. In order to fully understand this increasing interest in, and use of, SCM in construction, it is first necessary to understand the main factors that have been driving and shaping its development, not only in construction, but in a number of sectors of the economy. The two most important factors include the new environment within which business is increasingly being conducted, and the new thinking in relation to management approaches – in the case of the latter, not only within organisations but also externally in relationships with other organisations in supply chains.

Changes in the business environment

Many writers, from a wide range of perspectives, have been arguing for some years that the wider environment is changing significantly and that business is moving to a new paradigm. As discussed in Chapter 3, from the late 1970s new methods of production and ways of doing business began to emerge. The key ideas associated with the new emerging paradigm include:

- a significant shift towards empowerment to encourage continuous improvement through learning and innovation

- a change from dedicated mass-production systems towards more flexible systems that can accommodate a wider range of products, smaller batches and more frequent design changes – hence the 'economies of scale' are increasingly being replaced by 'economies of scope' and greater mass customisation
- a customer-focused approach with the customer brought into the centre of the business in order to align the organisation with the evolving needs and expectations of customers
- a move towards the greater integration of processes and systems within companies and between suppliers and customers, in order to align resources more closely to market and customer requirements and
- developing the appropriate organisational structure and culture to support and sustain learning and innovation.

It is the emergence of these new possibilities and challenges associated with the new paradigm that is driving organisations – including an increasing number in construction – to undertake a fundamental reassessment of the way in which they achieve their objectives through innovation, not only within their organisation but also in their networks of customers and suppliers. Another significant feature of the new paradigm is the growth of outsourcing and the increasing reliance on suppliers and subcontractors to build and maintain competitive advantage. Outsourcing refers to the activity of purchasing goods or services from external sources, as opposed to internal sourcing (either by internal production or by purchasing from a subsidiary of the organisation). This growing reliance on suppliers means that, for instance, a typical company can now spend an equivalent of 70% of their costs on the purchase of goods or services from other companies. This means that unless the quality, cost, design and delivery of inputs from their suppliers are appropriate, the final goods or service is unlikely to meet the needs of their end customers.

This increasingly important role of suppliers in achieving and sustaining competitive advantage was recognised more quickly in Japan than in other industrial countries. For instance, as a result of their rapid post-war growth and limited capital for investment, larger companies in the electronics and automotive industries subcontracted a larger proportion of their production. At this post-war time, however, there were very few good Japanese suppliers for the rapidly growing main industrial assemblers. The increasing beneficial role of subcontracting rested on the main assemblers' abilities both to control the process and rapidly improve the abilities of suppliers. The resulting supply system developed in Japan is not only vital to the viability of the main assembler or manufacturer, but can be seen as the basis of what is now being recognised as SCM.

During the 1980s and 1990s, there was also a notable increase in outsourcing in developed countries. The main motives for this included cost reduction (as external sources can enjoy greater economies of scale), access to specialist expertise, and greater concentration on an organisation's core competence by discarding peripheral operations. The potential disadvantages of outsourcing include reduced control over the operations involved, fragmentation, concerns over delivery and quality. SCM as a fifth generation innovation has been developed to address the problems associated with outsourcing, to introduce greater coordination and integration and to respond as effectively as possible to the needs of customers and changes in the wider environment.

Outsourcing in construction

With the increasing complexity of the design and specification of construction products and processes as described in Chapters 2 and 3, more finely subdivided engineering specialisms began to emerge to support the work of the architect. Also, general contractors were no longer able to undertake all the work or provide the substantial capital investment needed in the emerging specialisms as many of them required, for example, expensive plant and equipment. In addition, subcontracting has been giving general and main contractors the flexibility to deal with fluctuations in overall demand for construction products and services as they respond to the diverse needs of clients and their wide range of construction needs. It has also been argued that the greater use of subcontracting by general contractors during the period between the early 1950s to the early 1970s was a means of circumventing difficulties such as the poor relations that had developed with their workers (Ball, 1988). Subcontracting either pushed the labour relations problem onto another firm or removed the role of trades unions altogether, for example by employing workers on a labour-only, self-employment basis. All of these developments contributed to the fragmentation of the industry, the marginalisation of subcontractors and thus increased the need for more effective coordination and integration.

Origins of SCM

The term SCM began to be used in the early 1980s to refer to the management of material flows across functional boundaries within organisations. This innovation and others such as such as Just-in-Time (JIT) and Total Quality Management (TQM) often provided substantial internal improvements by breaking down barriers between departments and focusing on efficiency in managing core processes. With the increased reliance on suppliers as a result of increased outsourcing, SCM has been extended beyond the boundaries of a single business unit to comprise all those organisations and business units that have to interact in order to deliver a product or service to the end customer as illustrated in Fig. 7.1.

Similar to partnering, SCM derives from two roots of practically-oriented management theory: Operations Management and partnership philosophies. Within Operations Management a typical definition of a 'supply chain' as proposed by Christopher (1998) is:

> *a system whose constituent parts include material suppliers, production facilities, distribution and customers linked together via feed forward of materials and feedback flow of information.*

Figure 7.2 illustrates the origin of SCM in the context of Operations Management. These early internal and increasingly external improvement programmes led to the broader concept of SCM. This includes:

- purchasing and supply management
- physical distribution
- logistics
- materials management.

Figure 7.2 identifies the key management functions involved in the management of the supply chain for consumer goods manufacture. As can be seen, purchasing and supply management is concerned with the links with first-tier suppliers. More than merely buying, the aim is to obtain supplies according to the five 'rights':

- the right price
- the right quantity
- the right schedule
- the right quality and
- from the right source (Naylor, 2002).

It is now widely accepted that these features contribute significantly to the value offered by organisations in their own products and services to their customers. Establishing deep relationships with a few suppliers, so-called 'single sourcing', has resulted in improvements not available in traditional approaches to purchasing.

Physical distribution management (PDM) refers to managing the flows to first-tier customers. It deals with issues such as the number and location of depots, the type of transport and the scheduling of flows. The skills and investment required in this activity mean it is frequently outsourced to specialists. Logistics is usually seen as an extension of PDM in considering the flow of products to consumers. While PDM considers the

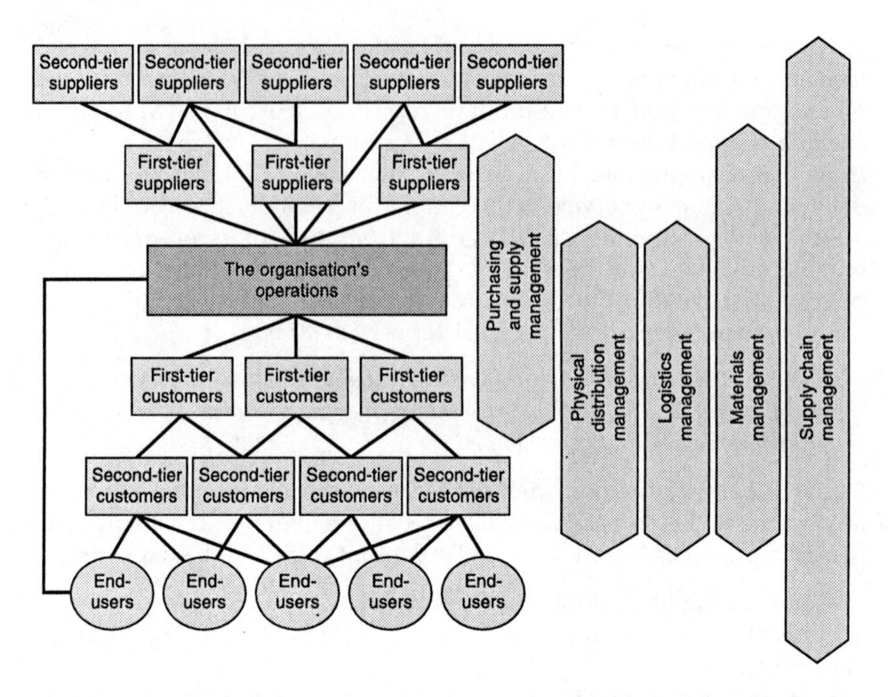

Fig. 7.2. Supply chain for consumer goods manufacturer (derived from Slack et al., *2001 and Hill, 2000)*

best way to deliver to the next tier, logistics seeks to optimise the whole chain (for example, to take joint decisions on packaging so that it not only protects the product during transit but also carries useful information and serves as a unit for display for the retailer). Materials management considers the flows of products both within and outside the organisation including many sites spread throughout different regions and countries in order to offer high efficiency and flexibility. It involves issues such as purchasing, location of plant and warehouses, stock control and design of transport systems.

SCM has emerged as a broad concept covering flows within and between organisations. It focuses on integrating all these functions, their processes and interfaces. This emergence of SCM as a fifth generation innovation has been driven not only by internal pressures to reduce costs and add value, improve efficiency and satisfy customers, but by a number of external factors which characterise the new paradigm such as globalisation, shifts in the nature of competition, systems development and ICT. Working more closely with suppliers requires high levels of information sharing, cooperation, increased openness and transparency which highlights the importance of the partnership philosophies root. This, as discussed in Chapter 6, has led to an increasing adoption of partnership approaches and inter-organisational collaboration to achieve significant mutual benefits involving sharing resources, information, learning and other assets (Mowery, 1988). As illustrated in Fig. 7.3, effectiveness in adopting partnership approaches is also linked to setting up the appropriate internal organisational and cultural changes.

The partnership philosophies emphasise the nature of the relationships between the organisations involved in the supply chain. The central concept is about building mutual competitive advantage through better buyer-seller relationships. At one end of the spectrum is the 'arms-length', or 'hands-off' contract for goods or services where a price is agreed for a completely specified, or a standard off-the-shelf, product or service. Apart from agreeing the product or service and the price, the buyer and supplier need know nothing of each other's processes and operations. In contrast to this approach, at the other end of the spectrum is a more involved and explicitly interdependent model based on a common purpose which is mutually beneficial leading to a sharing of profit and risk. This needs to take place in an atmosphere of trust based on sharing of information and knowledge in order to understand issues and problems as they emerge, and devise appropriate solutions. This often requires going well beyond the commitment of customers and suppliers associated with a formal contract.

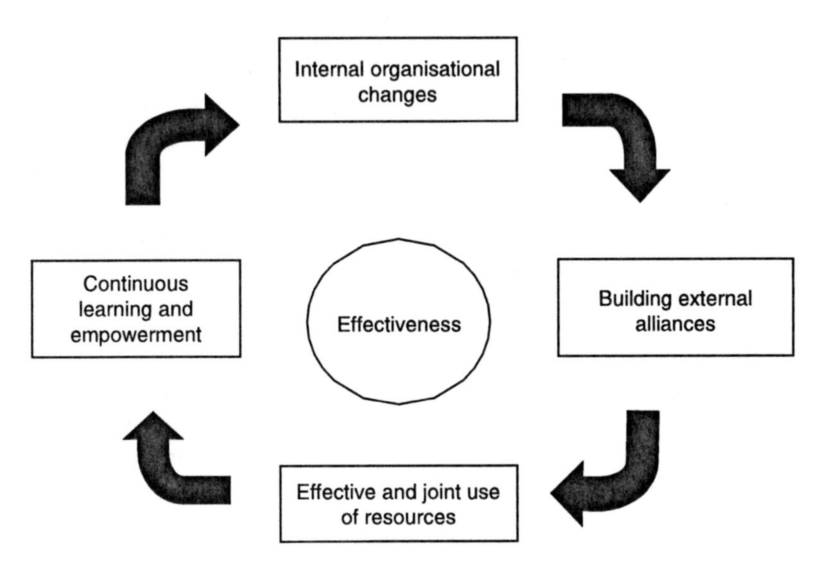

Fig. 7.3. The partnership root

Definition of SCM

As SCM is a relatively new concept, it is still in the process of being clearly defined. Christopher (1998) defines a supply chain as

the network of organisations that are involved through upstream and downstream linkages, in the different processes and activities that produce value in the form of products and services in the hand of the ultimate customer.

Most definitions describe supply chain management as the chain linking each element of the manufacturing and supply process from raw materials to the end-user, encompassing several organisational boundaries. This is well summarised by Christopher (1998) as

the management of upstream and downstream relationships with suppliers and customers to deliver superior customer value at less cost to the supply chain as a whole.

Most definitions link SCM with the integration of systems and processes within and between organisations, which include the upstream suppliers, and downstream customers. Accordingly, SCM can be seen as a set of practices aimed at managing and coordinating the whole supply chain from raw materials suppliers to the end consumer. It is also viewed as the coordination of manufacturing, logistics and material management functions across the organisations (Harland, 1996).

SCM is therefore associated with improvement programmes that have been broadened to include methods of reducing waste and adding value across the entire supply chain (Tan, 2000). The main objective is to develop greater synergy throughout the whole network of suppliers through a better integration of both upstream and downstream processes (New and Ramsay; 1997; Christopher and Jüttner, 2000). This significant emphasis on coordination and integration is strongly dependent on the development of more effective and longer-term relationships between buyers and suppliers with increased trust and commitment (Vollman *et al.*, 1997; Kosela, 1999). It is about adopting a more holistic approach in order to optimise the overall activities of companies working together to build greater mutual competitive advantage and greater customer focus. Fig. 7.4 shows a common way of visualising a supply chain.

However, this view is increasingly seen as an over-simplification compared to the model shown in Fig. 7.5. In reality, most companies have several if not scores of customers and suppliers. Often, several companies compete for the same customers and have common suppliers. A more realistic picture is more complex, with a multitude of relationships between customers and suppliers as shown in figure 7.5.

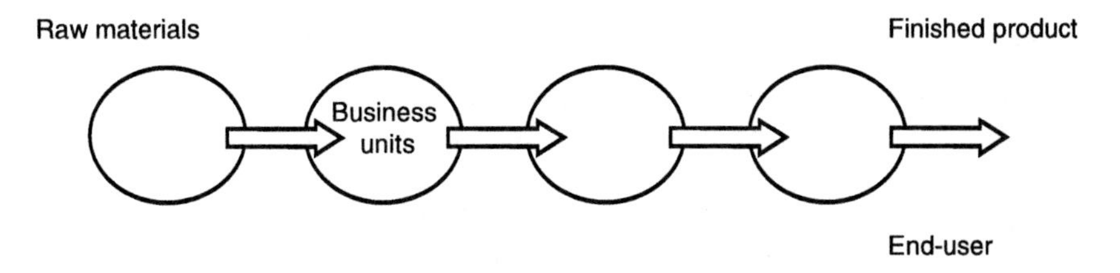

Fig. 7.4. A common view of a supply chain

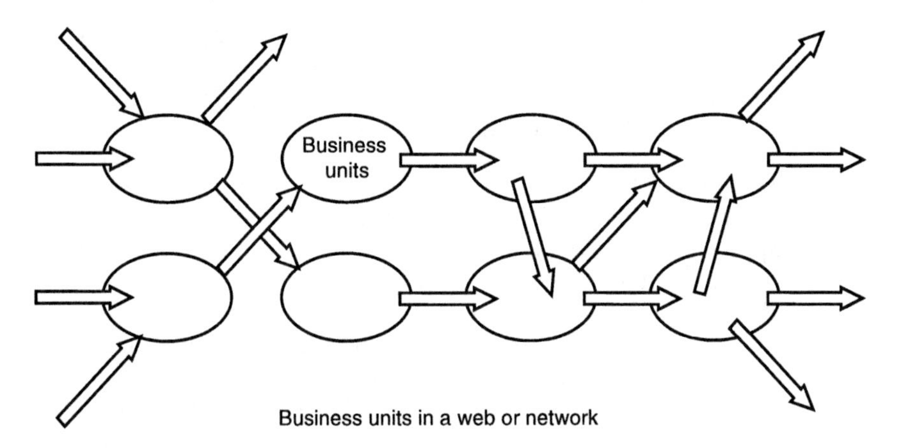

Fig. 7.5. Supplier networks

Key actions for implementing SCM

As already discussed in Chapter 4, the complexity of this innovation suggests that its implementation requires a number of actions which have to be considered concurrently rather than sequentially. It implies a process involving some degree of experimentation, learning and feedback mechanisms. It also needs to be contingent upon the organisations involved and their present relationships and wider environment. Hence, the implementation of SCM should not be seen as a linear process and the following features should not be seen as a sequential list of actions for implementation.

Agreeing a common purpose to ensure mutual competitive advantage

As with partnering, effective SCM requires a significantly higher level of joint strategy development where the members of a supply chain collectively agree a common purpose and jointly set strategic goals that are mutually beneficial. Most failures of SCM are mainly attributable to members of supply chains being in competition with each other and therefore not sharing their strategic thinking and not agreeing a common goal and strategy. Establishing a common purpose based on mutual advantage should reflect the degree of preparedness of the organisations and the scope of their relationship. In the first instance, this should focus on their development of contractual trust. A common purpose should be based on:

- reciprocity by which one is contractually obliged and subsequently morally obliged to give something in return for something received
- fair rates of exchange between costs and benefits and
- distributive justice through which all parties receive benefits that are proportional to their contribution and commitment.

Ensuring mutual benefit is perhaps the biggest challenge in removing hidden agendas and shifting from the often adversarial nature of buyer-supplier relationships to long-term and trusting relationships. Competitive advantage is seen as no longer residing only with a company's own innate capabilities, but also with the effectiveness of the relationships and linkages that the firm can forge with other organisations in the supply chain. The new dynamics of these supply chain relationships are built on the fundamental idea of sharing both profit and risk. In this context it is no longer appropriate to view suppliers and customers as though they are independent entities competing with each other.

Developing more collaborative inter-organisational relationships

Evidence shows that long-term agreements can improve relationships and reduce conflict and transaction costs. Instead of pursuing short-term contracts characterised by frequent bidding and switching costs and the costs of pursuing claims and resolving conflicts, purchasers and suppliers can direct their efforts towards value-adding activities. Long-term purchase agreements are a prerequisite to achieving ongoing and closer cooperation between purchaser and supplier leading to, for example, early involvement of the supplier in the design of new products, services and processes as illustrated in the HBG case study.

Case study 7.1: HBG – improving performance through ongoing supplier relationships

Between 1997 and 2000, HBG's North-East office used continuous improvement processes to hand over nine cinema shells in an increasingly short time. In 1997, the contractor completed the on-site construction of the shell-and-core for a nine-screen Virgin multiplex cinema at Boldon, near Sunderland, in 38 weeks. A year later, the team completed a nine-screen UCI complex at Silverlink, North Shields, in 34 weeks. In 1999, the contractor took 30 weeks to complete the nine-screen Virgin multiplex cinema at Kingswood, Hull. By the time the team started on the Hull scheme, it had worked on nine cinema projects. In addition to cutting time on site, the fact that the subcontractors had built relationships was also starting to reduce the lead-in before construction started on site.

Part of the reason for this reduction was the project team's familiarity with many of the construction details. However, it was mostly attributed to the partnership that HBG had formed with Barrett Steel Buildings, which meant that a detailed steelwork design was being produced early in the project. The steelwork subcontractor was prepared to undertake detailed design work to price the scheme, safe in the knowledge that if HBG won the job, it would as well. Partnering arrangements with the key subcontractors over a series of cinema projects allowed the team to standardise and understand the way key components fitted together.

Andy Pearson, The Benchmark series, *Building*, 21st January 2000.

SCM which is dependent on win–win thinking permeating the culture of the supply chain can help overcome the often deeply ingrained adversarial attitudes. This involves the two-way exchange of information and the building of closer, long-term relationships so that a supplier may have a guarantee of business for several years, rather than having to re-bid for each order. In return, the supplier is required to hold or reduce prices, improve quality and delivery, and introduce more efficient processes. The use of open-book accounting can also lead to more transparency and open communication between the members of the supply chain. Often, the buying company will also commit itself to helping the supplier to improve its competencies through supplier development activities.

Selecting the right partners

The move to close, long-term and more collaborative customer–supplier relationships has increased the importance of choosing the most appropriate partners. It requires both customers and suppliers to have an overall understanding of the nature of their businesses and their links with supply chains. In this complex setting, companies need to realise that some relationships between customers and suppliers will be more important than others in terms of their own competitive advantage. Increasingly, companies are also looking at a wider set of competencies to guide their selection of their closest partners. Vollman *et al.* (1997) have identified three different types of competencies:

- *distinctive* competencies that provide the organisation with a unique competitive advantage
- *essential* competencies that are vital to the effectiveness of the organisation at the operational level and
- *plain* competencies that have no direct effect on the product or service delivered.

Reducing and shaping the supplier bases

Developing closer customer–supplier relationships requires considerable time and resources. This explains the need to reduce the number of suppliers significantly (often from several thousand to less than a hundred) in the supply chain. This smaller number of suppliers are then often organised into a hierarchy of tiers, as shown in Fig. 7.6, which reflects the importance and competencies of the suppliers. This

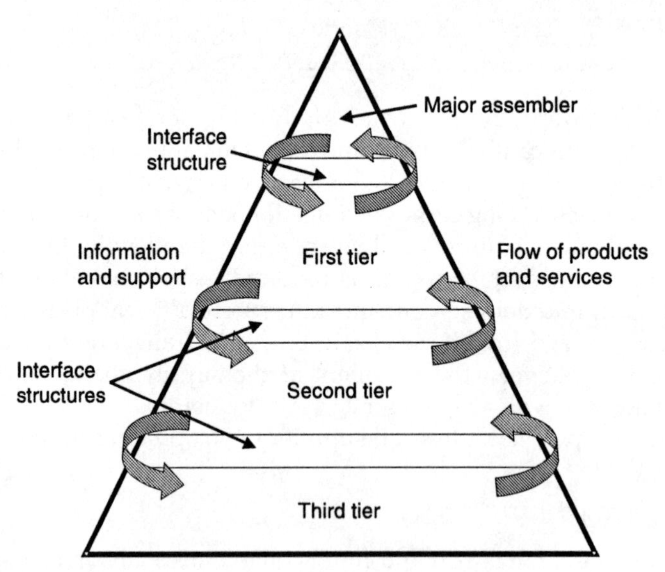

Fig. 7.6. Pyramidal supply system

rationalisation of the supplier and customer bases allows organisations to focus on their most significant customers and suppliers.

This hierarchy, which often takes the form of a pyramid, demonstrates the tiering commonly found in the automotive industry. In this example, there are multiple layers or tiers roughly delineated by the size of the firms and their roles in the supply chain. At the apex of the pyramid sits the final assembler who is supplied with sub-assemblies by first tier suppliers. In turn, these first tier companies are supplied by a larger number of second tier suppliers. These second tier suppliers have their own subcontractors who provide them with specialist process abilities. In some instances there may even be fourth and fifth tier suppliers (Hines, 1994).

Managing the interfaces between the organisations involved

Managing the interfaces between customers and suppliers, based on closer inter-organisational relationships, is a critical element of SCM. The effective management of this relationship requires that the organisations have the appropriate internal culture and organisation, as summarised in Fig. 7.3. This facilitates the sharing of information and learning in order to achieve a greater coordination and integration of processes across the interfaces.

An advanced form of interface structure between a customer and a supplier is where organisational boundaries become blurred, as shown in Fig. 7.7. This interface structure allows for both joint strategy development and day-to-day problem solving and operational activities. Joint teams can also be formed to cooperate on specific issues such as innovation, R&D, market environment, sustainability and end-user satisfaction.

Shifting from inter-firm competition to network competition

In SCM, organisations seek to make the supply chain as a whole more competitive through the value it adds and the costs it reduces overall. Real competition is not between organisations but rather between supply chains. Reducing costs and increasing value are both important parts of gaining competitive advantage. The traditional approach sought to achieve cost reductions or profit improvement at the expense of other participants in the supply chain. Increasingly, however, organisations are realising that simply transferring costs upstream or downstream does not make them any more competitive as, ultimately, all costs will make their way to the end customer. The shift from classic inter-firm competition to supply chain competition is illustrated in Fig. 7.8.

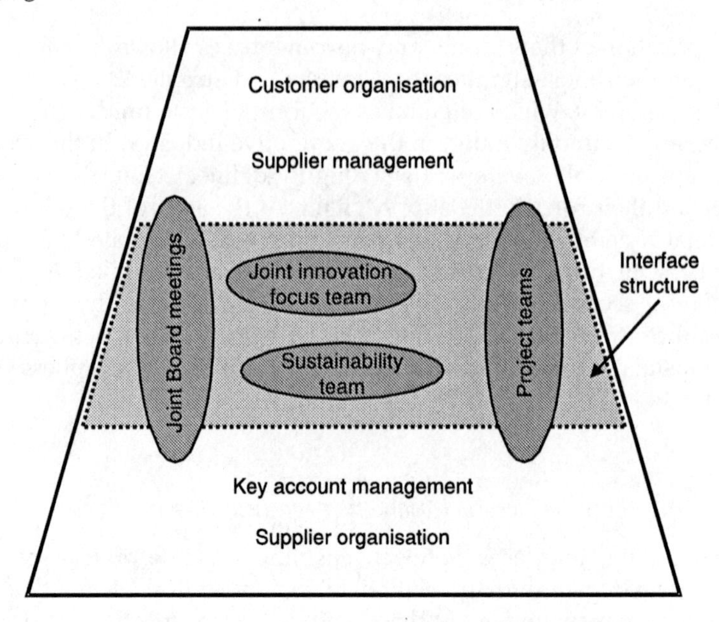

Fig. 7.7. An advanced interface structure (derived from Christopher and Jüttner, 2000)

Performance improvement requirements and management

Continuous improvement and performance measurements are key elements of SCM. Robust measures of performance and targets for improvement are also necessary to prevent the danger of complacency developing in long-term relationships. Specifically, organisations need measurement systems to identify:

- supplier performance and improvement opportunities
- performance trends
- appropriate suppliers
- resources needed for supplier development and
- the overall effectiveness of supply chain management in improving performance.

Supplier development

Empowerment through supplier development is another key factor of SCM and its ability to unlock additional value and competitive advantage from suppliers. Supplier (or customer) development is where a partner in a relationship modifies or influences the behaviour of the other partner with a view to increasing mutual benefit. Cross-functional

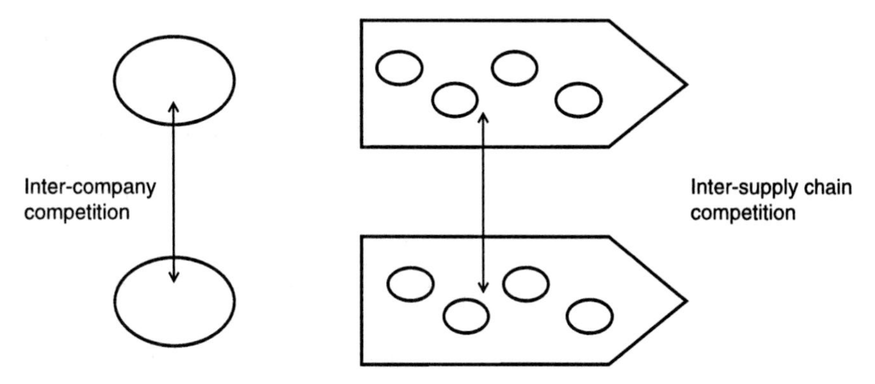

Fig. 7.8. The shift to supply chain competition

teams from the organisations work closely to share learning, solve problems and seek improvements in their internal and interface processes.

Governance

It is important to understand power in the context of SCM. In supply chains it can be seen as the ability of one individual or organisation to influence the behaviour of another individual or organisation in order to achieve their desired situation and outcome. To have power is relative not absolute, since it is contingent upon the context and relationships within the supply chain. There are a number of types of power in supply chains:

- the power to reward which is related to the extent that one individual, group or organisation can reward another in the supply chain
- the power to coerce which is determined by the ability of an individual, group or organisation to sanction or punish others for unwanted behaviour
- the possession of legitimate power which corresponds to the extent to which an individual, group or firm feels that it is right for another individual, group or organisation to take action and exercise their power
- the expert power which an individual, group or firm perceives another to have in relation to key knowledge or specialised technical skills
- the referent power which corresponds to the extent to which others in an organisation or supply chain wish to identify with a single

individual, group or organisation on the basis of, for example, their leadership style, position, or approaches in dealing with a difficult issue

- the power arising from an individual or group's ability to cope with uncertainty and ambiguity: in creating certainty for others in the supply chain individuals, groups and organisations become more powerful
- the power arising from indispensability which an individual, group or organisation has if it cannot be substituted in the supply chain and
- the power from relationships and centrality which is conferred by being well-networked into the whole supply chain.

The more types of power that an individual or group of organisations has, the greater their influence in the supply chain. For example, a supplier which is indispensable from the perspective of the customer will be more powerful and will have potentially more influence than an undifferentiated supplier. Key customers who buy large volumes of products or services on a regular basis are also more likely to have greater power over their supply chains than those who are small or only occasional customers.

It is important to understand how the exercise of power in supply chains can determine whether the relationships between suppliers and customers are either loosely or tightly coupled. For example, if supply relationships are tightly coupled they may impinge on the operation of another firm in the network and paralyse its strategic actions. On the other hand, if relationships are too loosely coupled it can affect the management of the interfaces and the overall effectiveness of the supply chain.

This review suggests that the successful implementation of SCM is associated with the following key actions:

- choosing the best suppliers and integrating them into a rationalised supply base
- developing long-term, close, stable, win–win and trusting relationships
- breaking down the barriers between internal departments, business processes and between companies to improve the management of interfaces
- coordinating the supply chain to manage fluctuations in demand and provide more predictability
- sharing information, learning and resources to develop the capacities and competencies of the whole supply chain

Potential benefits of SCM

A review of literature in a number of sectors of the economy provides evidence of the significant benefits to be gained from the synergies developed from the successful implementation of SCM. Sako *et al.* (1994) suggest that, for instance, closer collaboration in European vehicles manufacturing has helped speed up the rate of performance of improvements in order to meet global competition. The success of the Japanese car manufacturers in introducing new models faster and with fewer labour hours has been attributed to the early, and more proactive involvement, of their suppliers in product design (Lamming; 1993). Shorter lead times, reduced total cycle time and smoother and more responsive flow of materials and products are also achieved through an effective management of interfaces of the entire supply chain (Hines, 1994 and Womack and Jones, 1996). Towill (1996) argues that time compression has a had a major impact on the accuracy of demand forecasting, the time taken to detect defects, the time to bring new products to markets, and the amount of work in progress. He also identifies a number of other benefits including:

- elimination (removing a process)
- compression (removing time within a process)
- integration (changing interfaces between processes) and
- concurrency (operating processes in parallel).

This greater collaboration between organisations within the supply chain can also result in increased internal and external customer focus (Lipparini and Sobrero, 1994). The collaborative SCM approach, based on frequent and direct communications between suppliers and customers, can help in defining and agreeing the customer requirements and the means by which they can be satisfied. These new types of relationships are increasingly perceived as a means to utilise resources better throughout the whole supply chain (Dubois and Gadde, 2000).

There are examples of where SCM is delivering significant performance improvements and increased competitive advantage (Burgess, 1998). It can also be an important element in innovation in products, processes and organisation (Holti, 1997) as information can be more readily shared and knowledge identified, captured and disseminated throughout the organisations in the supply chain (Edum-Fotwe *et al.*, 2001). This sharing of information and knowledge has led to joint learning resulting in significant improvements in a number of aspects of the supply chain. This shared learning can also lead to better problem identification and joint solving, and allows the more ready application of techniques such as value management, analysis and engineering. It can also improve the predictability of changes in the external environment as a result of more effective communication and greater synergy. Greater compatibility and integration of processes and systems achieved through collaboration and partnership within the supply chain are more likely to lead to increased flexibility and more responsiveness to changes in the external environment. All these potential benefits are linked to the development of trust through greater mutual understanding, transparency in transactions and commitment (Ali *et al.*, 1997).

Increased trust

The concept of trust is very complex, not clearly defined and interpreted differently by the literature (Smeltzer, 1997). Most of the literature relates trust to predictability, risk, vulnerability, cooperation, personal traits and confidence. Trust is, on the whole, associated with the confidence that a partner will produce a mutual beneficial behaviour. This view challenges the prevailing view of economic theory that organisations and individuals are self-interested and will pursue their interests opportunistically and often with guile. Trust is based on the belief that a partner is reliable and will fulfil the perceived obligations of the relationship. This leads to a reduction of the risk that partners will not perform an action detrimental to the relationship. Trust is also defined in terms of good intentions and confidence (Cook and Wall, 1980). There is a clear connection between predictability of behaviour and trust, both of which are seen as means to manage and reduce uncertainty. However, it must be recognised that risk can be increased in the early stages of building closer relationships as partners increase their vulnerability as a precondition to the formation of trusting behaviour (Kellerman, 1987). Trust is therefore related to a degree of expectation which, if not fulfilled, may lead to disappointment. This implies the need to recognise and accept that risk exists.

When trust is low, other control mechanisms need to be employed. Typically, legalistic remedies (e.g. accrediting organisations, insurance, bonds, guarantees, etc.) are used either to compensate for the lack of trust in exchange, or to create the conditions under which trust might be restored. However, these kinds of mechanisms can be counterproductive and lead to even higher levels of mistrust. In addition, these remedies to high levels of mistrust are costly and impede performance. When trust is present within and between organisations the cost of transactions can be lower as fewer controls are needed to measure, monitor and control performance. Forecasts of future events are more realistic, as are budget projections. Cooperation improves along the value chain, productivity improves, profitability is enhanced and the sharing of learning and innovation is encouraged.

It is also important to recognise that trust must go beyond predictability if it is to produce effective and lasting relationships and partnerships based not only on contractual and competence trust but also in the longer term on goodwill trust (Sako,1992). Contractual trust emerges when each partner will follow written or oral contracts. Competence trust refers to the ability of the partner to complete tasks to a given standard while goodwill trust corresponds to the situation where a partner, driven by mutual benefits, does more than expected and goes beyond predictability and the agreements of a contract. As illustrated in Fig. 7.9, trust is cumulative and takes time to build. Contractual trust needs to be established prior to competence trust which in turn will lead to the development of goodwill trust. The latter can continue as the partnership and the relationships develop.

What distinguishes 'goodwill trust' from 'contractual trust' is the expectation in the former case that partners are committed to take initiatives (or exercise discretion) to exploit new opportunities over and above what was explicitly promised. The key difference here is that partners are not only looking after their own interests but are also seeking to offer their partner a competitive advantage. If both partners do the same, then the combined efforts of both customer and supplier will lead to a mutual competitive advantage, which will help in selling their collective product or service to the end-user.

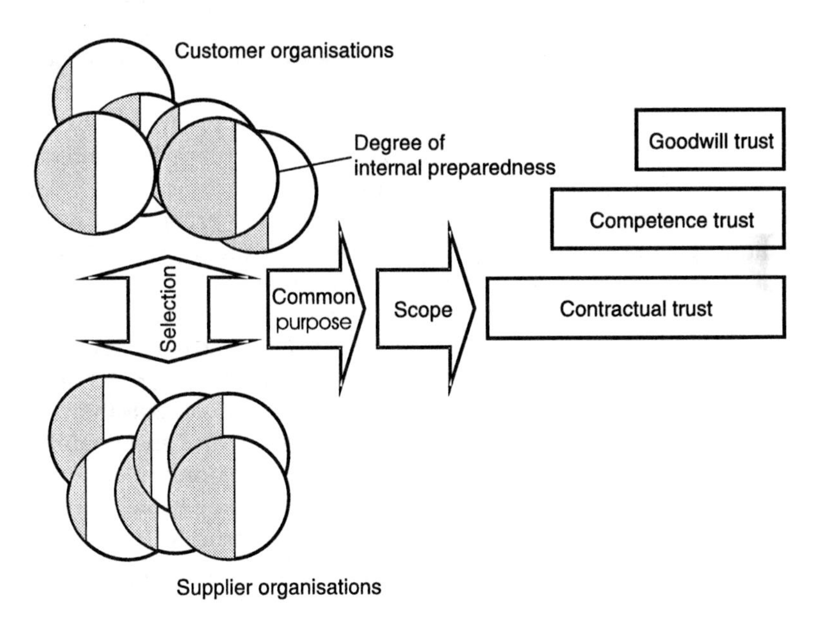

Fig. 7.9. Development of trust

However, it is unrealistic to describe the development of trust as a linear process, as the three types of trust are interlinked. In general, contractual trust is often the starting point for the development of trust. Competence trust cannot develop without contractual trust and goodwill trust cannot exist if the former two are not present. It is a long process and takes time and resources to develop.

Although the literature identifies trust at the individual, organisational and collective level, there is consensus that the common denominator or rather the starting point is indeed individual or interpersonal trust. Interpersonal traits are considered as forming the foundation of trust (Dasgupta, 1988; Farris *et al.*, 1973). Interpersonal trust is associated with the level of expectation that the word, promise, verbal or written statement of another individual or group can be relied upon (Rotter, 1967). However, the propensity of groups and individuals to trust others is often determined by different developmental experiences, personality types and cultural backgrounds (Hofstede, 1980).

Ongoing interaction allows individuals to know and understand each other and develop higher levels of trust. This is why Kanter and Myers (1989) claim that personal relationships and personal connections are crucial in forming effective relationships and greater trust between groups or organisations. James and Saad (1998) argue that these personal and often informal relationships develop trust by exerting pressure for conformity to expectations. These interpersonal and informal ties can increase over time and lead to greater trust in managing partnerships. In such situations, where trust emerges as a consequence of effective and informal relationships rather than a prerequisite to the relationship, informal contracts are used rather than adversarial, detailed and formal contracts. It is clear that trust between organisations is progressively and incrementally built as organisations and individuals repeatedly interact (Good, 1988). Repeated personal interactions across firms can encourage higher levels of courtesy and consideration while the prospect of ostracism among peers can discourage opportunism. Mutual expectation of repeated trading over the long run is a major incentive for cooperation and effective partnerships (Axelrod, 1984; Kreps, 1990).

There is, however, a risk associated with predicting behaviour and sustaining trust merely through informal relationships. A wider level of trust needs to be achieved within and between the organisations in order to ensure an acceptable degree of predictability and sustainability in the relationship. Emphasis needs, therefore, to be placed on organisational traits, learning and culture which can be conducive to an organisations' identity, image and reputation upon which trust can be based (Smeltzer, 1997). This can be embedded within the relationship through a clear common purpose, mutual objectives, openness, transparency and the sharing of information and knowledge. Hence, internal preparedness based on a culture of trust can be seen as a vital prerequisite for effective and sustainable inter-organisational relationships.

As shown in Fig 7.9, the degree of internal preparedness, the selection of customers and suppliers and the degree of common purpose should help determine the type and nature of trust to be achieved in the relationship. In the case of contractual trust this can be based on a contract which sets out the expected behaviour of each party. Such contractual trust should lead to the elimination of disputes, conflict and adversarial relationships. Improved relationships coupled with more appropriate internal preparedness should lead to more sharing of information, greater joint problem solving, joint learning and innovation and competence trust. Feedback on the behaviour of the individuals and organisations involved will influence the scope of the common purpose. Negative feedback may result in reverting back to contractual trust or even opportunistic behaviour, while positive outcomes would take the organisations towards the next step on the journey: goodwill trust. Achieving goodwill trust is a long process which evolves slowly and is based on a continuous improvement approach.

Concerns associated with SCM

As in all innovations, SCM has some significant weaknesses and inhibitors to its successful adoption and implementation. It is a long, complex and dynamic process and its implementation requires a thorough understanding of the concept (Akintoye *et al.*, 2000; Whipple and Frankel, 2000; Edum-Fotwe *et al.*, 2001). It is also seen as closely dependent on the ability to create, manage and reshape relationships between individuals and organisations within the supply chain (New and Ramsay, 1997; Spekman *et al.*, 1998; Harland *et al.*, 1999). The implementation of SCM requires new intra- and inter-organisational arrangements and culture all of which require considerable commitment and resources and take time to develop. Time is required for learning, achieving mutual understanding, agreeing a common purpose, selecting the appropriate customers and suppliers, building trust, negotiating objectives and coordinating activities. Difficulties which can emerge include multiple and often hidden goals, power imbalances, different cultures and procedures, incompatible collaborative capability, the tension between autonomy and accountability, over-dependence, complacency leading to a reduction of long-term competitiveness, and continuing lack of openness and opportunistic behaviour (Huxham, 1996; Cox and Townsend, 1998). It needs to be recognised that over-tight networks can reduce agility and responsiveness to new market conditions and opportunities and lead to the risk of contravening legislation aimed at promoting competition. They can also result in having to continually pass inappropriate levels of value to the customer and missing out on technological and organisational innovations in other supply networks.

There is an emerging recognition that a major difficulty associated with SCM is what constitutes effective relationships and how their effectiveness can be assessed. SCM is based on soft or less tangible aspects which current metrics of performance are not adequate to measure. The main tools currently include the Business Excellence model, Competitive Positioning Matrices and the Partnership Sourcing model. A number of aspects of the Business Excellence approach relate specifically to customer-supplier relationships as shown in Table 7.1.

Barriers to be overcome in implementing SCM

A number of barriers to the successful implementation of SCM are attributable mainly to insufficient internal preparedness by the participating organisations. These include:

- lack of commitment to SCM
- insufficient understanding of the concept of SCM
- lack of strategic leadership
- lack of understanding of the new model of supply chain competition
- inappropriate organisational structure
- lack of leverage by any of the members of the supply chain to affect change and modify behaviour
- unwillingness to adopt win–win thinking
- insufficient allocation of time and resources to build internal and external relationships
- lack of common purpose and transparent and mutually beneficial goals
- resistance to the sharing of information, procedures and processes
- inappropriate distribution of risk
- inappropriate exercise of power and
- lack of commitment to innovation and learning.

As can be seen from the above review, SCM demonstrates the key features of a 'fifth generation innovation'. It is a multi-factor process which involves different functions, stakeholders and variables, and a whole sequence of events. It is a complex, dynamic and long process which involves individuals and groups from within and between organisations. SCM has shifted the emphasis from internal structure to external linkages and processes, and is dependent on the interaction between the organisation and its external environment, with strong

Table 7.1. Aspects of the Business Excellence model relevant to SCM

Leadership
How the behaviour and actions of the executive team and all other leaders inspire, support and promote a culture of TQM.
How leaders are involved with customers, suppliers and other external organisations.
Areas to address *could* include *how* leaders:

- meet, understand and respond to needs
- establish and participate in partnerships
- establish and participate in joint improvement activities
- actively participate in professional bodies, conferences and seminars
- promote and support TQM outside the organisation.

How leaders recognise and appreciate people's efforts and achievements.
Areas to address *could* include *how* leaders are involved in recognising in a timely and appropriate way:

- individuals and teams outside the organisation (e.g. customers, suppliers).

Policy and strategy
How the organisation formulates, deploys, reviews and turns policy and strategy into plans and actions.
How policy and strategy are based on information which is relevant and comprehensive.
Areas to be addressed *could* include *how* the organisation uses information relating to:

- customers and suppliers.

Resources
How the organisation manages resources effectively and efficiently.
How supplier relationships and materials are managed.
Areas to address *could* include *how* the organisation:

- develops supplier relationships in line with policy and strategy
- maximises the added value of suppliers
- improves the supply chain
- optimises material inventories
- reduces consumption of utilities
- reduces and recycles waste
- conserves global non-renewable resources
- reduces any adverse global impact of its products and services.

(*continued*)

Table 7.1. (contd.)

Processes
How the organisation identifies, manages, reviews and improves its processes.
How processes are systematically managed.
Areas to address *could* include *how* the organisation:

- resolves interface issues inside the organisation and with external partners.

How processes are reviewed and targets are set for improvement.
Areas to address *could* include *how* the organisation:

- uses information from employees, customers, suppliers, other stakeholders, competitors and society, and data from benchmarking in setting standards of operation, priorities and targets for improvement.

How processes are improved using innovation and creativity.
Areas to address *could* include *how* the organisation:

- uses feedback from customers and suppliers and other stakeholders to stimulate innovation and creativity in process management.

---

feedback linkages and collective learning. Its success is associated with the challenging and difficult development of a new culture based on long-term and closer intra- and inter-organisational relationships, shared learning, greater transparency and trust. Intra-organisational preparedness is an important prerequisite for the effective implementation of an SCM strategy.

As the main objective of SCM is to increase mutual competitive advantage through improved relationships, integrated processes and increased customer focus, it may well be relevant to construction with its adversarial relationships, fragmented processes and lack of internal and external customer focus. However, contingency theory suggests that there may well be specific issues related to implementing SCM in the context and culture of construction.

The emergence of SCM in construction

As was seen in Chapters 2 and 3, construction has a long history of introducing changes aimed at improving its performance. Partnering, for example, as examined in Chapter 6, is a recent key change in the shift from fragmentation and short-term adversarial relationships to greater integration and longer-term inter-organisational relationships. This has led to a growing interest in introducing SCM in construction to further integrate processes, manage interfaces, reduce uncertainty and increase overall effectiveness in a greater part of the supply chain. The aim of this section is to evaluate the relevance, adoption and implementation of SCM in construction. It seeks to address whether construction has the ability to use and sustain SCM between its large number of discrete work packages and across organisational boundaries. This section also highlights the prerequisites and the key characteristics of construction which need to be taken into account for a successful implementation of this innovation.

As discussed in previous chapters, over the past 50 years or so there has been a succession of reports into the state of the UK construction industry and there have been many calls for action to improve its performance and competitiveness. An analysis of these reports indicates that the problems facing construction can be categorised into three broad areas:

- poor relationships
- insufficient integration of processes and
- a lack of customer focus, as shown in Fig. 7.10.

A number of problems in construction's supply chains downstream of the main or general contractor (Jones and Saad, 1998) are outlined in Table 7.2.

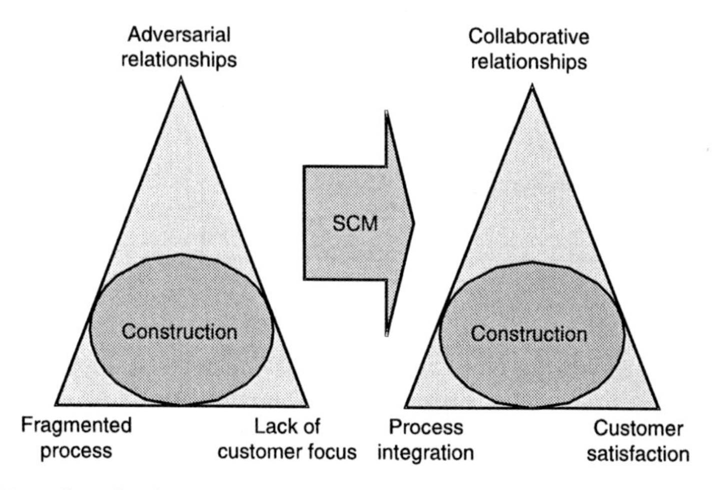

Fig. 7.10. The role of SCM in addressing the key problems in construction

Table 7.2. The main problems of construction

Relationships	Processes	Customer focus
Lack of trust leading to conflict	Fragmented nature of the design process	Fragmented nature of the construction process and poorly integrated value chain
Onerous contract conditions and unfair loading of risk	Inadequate design period	Insufficient focus on internal and external customer requirements
Lack of understanding of the risks involved and their consequences	Late, poor and incomplete design information lacking specialist contractor input	
Unfair selection procedures	Poor overall planning with inadequate lead-in time	Insufficient understanding of specialist contractors' requirements
Unfair payment procedures	Fluctuations in demand for the products and services of the project participants	Unclear statements of requirements and ambiguous project information and tender packages
Perceived poor status of specialist contractors	Failure to involve specialist contractors early enough in the process	
Failure to communicate with specialist contractors and to view them as equal project partners		

SCM, with its strong emphasis on improving relationships, a process-oriented approach and increasing customer focus, is an appropriate strategy for improvement in construction as shown in the Mace case study.

Case study 7.2: Mace – developing greater customer focus

As clients become more demanding and move away from cheapest-price tendering to partnering, construction organisations are paying more attention to winning clients, giving them what they want and, crucially, keeping them in a long-term relationship. Measuring customer satisfaction is one way to ensure clients are retained. A number of organisations such as construction manager Mace have realised that ticking boxes on a form is not good enough if they want to find out what the customer really thinks. As an alternative they are appointing people who are independent of the project team to interview the client on a one-to-one basis at various stages of the building process. However, it is how construction organisations use this information to improve their service that will determine whether customers are satisfied.

An industry leader in the use of key performance indicators, Mace's approach is to employ a full-time best-practice manager who carries out three one-hour-long interviews with the client for each project. This gives customers a chance to have their say. If something has gone wrong in the project they can find out why, or they can confirm that everything is going fine. The first interview takes place at the start of the project to establish customer expectations, the second is held halfway through to find out how Mace is performing, and the final interview is carried out on handover to determine overall customer satisfaction. Mace conducts a further interview at the end of the defects liability period to ensure quality is maintained.

In each interview the client is taken through 47 questions covering 5 headline topics. These include the following points:

- how Mace personnel have performed on delivery, relationships, and health and safety
- how effectively the programme was planned and managed
- how the budget and suppliers were managed and
- quality control.

The interviewer asks the client for specific examples of practice, and explains why each question has been given a particular rating. A report is then prepared that is seen only by the project team, plus Mace directors. This information then feeds back into improving site practice.

Thomas Lane, The Benchmark series, *Building*, 27th September 2002.

There is now a growing interest in SCM as an innovation to address the problems impeding construction's performance, and to tackle the issues of uncertainty and interdependence. This is clearly confirmed by a survey on SCM[1] undertaken by the authors. SCM, as shown in Figs 7.12 and 7.17, is perceived as an innovation which can help construction to integrate its processes, enhance and develop longer-term relationships, and create greater value.

This is being led, as suggested in Fig. 7.11, by more informed private sector clients who adopted partnering in the early 1990s in their attempt both to increase the degree of collaboration between their preferred consultants and contractors and to extend this approach downstream to include key specialist and trade subcontractors and suppliers. Figure 7.12 and the Shepherd case study show that main contractors are also playing a key role in implementation of SCM through a greater integration of both upstream and downstream participants and processes.

Case study 7.3: Shepherd – building repeat business through SCM

Repeat business is so important for construction organisations. This means having to continually learn how to provide new services and align them with new clients who are looking for those services. Shepherd's approach includes the following aspects.

Customer strategy – recognising niche markets – in this case educational and research facilities where Shepherd had existing expertise. Knowing what the customer is likely to need and demonstrating a thorough grasp of the brief are essential skills that gain customers' confidence.

Build the supply chain – this involved culling poorer performing suppliers from a database of 26 000 companies. Shepherd awarded 'Approved' status to some 250 of its best suppliers. These attended workshops and signed charters to show they were committed. Shepherd then selected only one in five Approved suppliers for further elevation to 'Preferred' status. This involved evaluation of questionnaires completed by the suppliers and visits to their premises.

Creating best value – Shepherd hold value workshops during the detail design phase of their design and build contracts. They believe that the expertise and creativity that is often latent in trade

[1] The aim of this survey was to capture general information about construction organisations adopting SCM, their perception and expectations, and the reality of SCM in construction. The sample of respondents comprised 110 organisations representing the main participants in the development process.

specialists can be drawn out in a stimulating forum. An example in Cambridge came from NG Bailey who developed the prefabricated plastic pipe assembly first used in phase one of the project. They found a supply of valves that doubled as connectors between the plastic trunk system and the copper branches. A *Rethinking Construction* team member visiting the demonstration project helped Shepherd to explain the best value objective to suppliers.

Measuring performance – many of the initiatives implemented, such as risk and value workshops, approved supplier workshops and preferred supplier visits are now being formally monitored. Shepherd's set of 13 KPIs includes industry standards such as customer satisfaction, cost and time predictability and safety. They also have bespoke indicators that measure rate of tender conversion, subcontractor performance, physical waste, utility usage and supply chain expenditure. Their indicators for voluntary leavers and training provision pre-empted the People Performance Indicators promoted by Rethinking Construction's Respect for People steering group.

One of the major problems was that project personnel did not at first fully understand the KPI initiative and this made data collection more difficult than it should have been. It was necessary to explain the process and purpose of measuring KPIs and also demonstrate that the indicators were practical measures of how the project and business were performing. People needed to know that it was not an academic exercise, but had the potential for changes that would really give the company a competitive edge.

Future developments of the strategy include generating more consistent data for KPIs which has, to date, proven difficult to collect and extract from the company's data system. Shepherd have also identified the benefits of on-going training in collecting data and developing in-house software to ease data extraction and reporting. The KPIs have provided results that point to trends and opportunities for improvement. Complacency must be avoided as it is tempting to stop looking for ways to improve when improvements become more difficult.

Movement for Innovation, Case History, www.m4i.org.uk

The results of the survey suggest that most respondents perceive SCM as a multi-factor process built around closer intra- and inter-organisational relationships, with over 92% of the respondents believing that 'long term and stable relationships' are 'important' or 'very important' features of SCM. Only 11% of respondents are involved in strategic part-

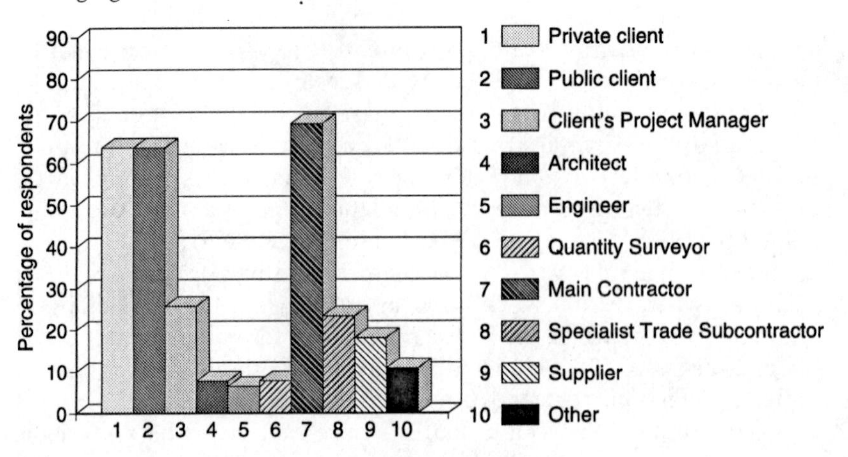

Fig. 7.11. Drivers of SCM in construction

nering (9%) or Prime Contracting (2%) which are the only procurement approaches allowing formal and long-term relationships. This demonstrates a present reluctance to develop the type of relationships associated with SCM. It also suggests that the overwhelming majority of relationships in construction are still mainly short-term and contractual. The survey also indicates a reluctance to reduce the number of suppliers and customers, negotiate clearer common objectives, openly exchange data and information and share learning and innovation. This reflects the industry's resistance to fully embrace a new culture which is needed for an effective implementation of SCM relationships, as shown in Fig. 7.12.

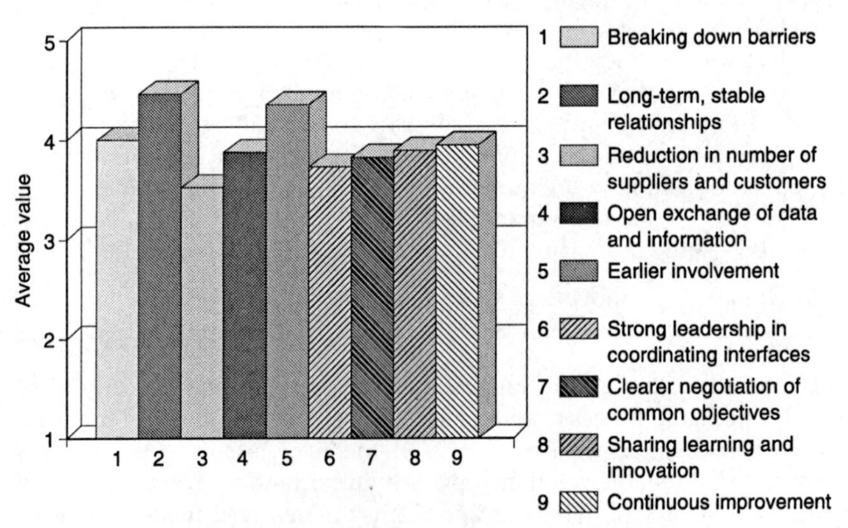

Fig. 7.12. The key features of SCM in construction

Encouraged by improvements brought about through partnering, these pioneers of more collaborative approaches have continued to develop closer relationships with their partners and to further integrate project and supply chain processes. A number of these clients and their consultant and main contractor partners are also beginning to extend the adoption of longer-term and more collaborative relationships downstream of the main contractor, to include key specialist and trade subcontractors and materials and component suppliers. Using their leverage in their supply chains, these frequent users of construction services have been able to make the transition successfully from project-specific partnering, through strategic partnering, and on to SCM as shown in Fig. 7.13.

With the encouragement provided from a number of reports into the inadequacies of construction procurement in the public sector, HM Treasury and the Office of Government Commerce (OGC) have been advocating partnering and SCM as ways of delivering value for money. Some public-sector clients, such as Devon Council, have responded by attempting to build the new purchaser-supplier relationships associated with SCM into their procurement of construction products and services.

Case Study 7.4: Devon County Council - setting up framework partnerships ensuring Best Value and probity

In 2000/2001 Devon County Council undertook a review of its property services department – which provides a full range of property services to client departments within the Council. The review made a number a recommendations including the need to produce a business case to look at the future position of the organisation within the council and to consider more innovative means of procuring construction and maintenance following the lead set by 'Rethinking Construction'.

With the assistance of consultants from CIPFA a business case was made for procuring framework agreements to cover construction (broadly new capital works) and maintenance (including minor new works). The scale of procurement dictated that the European (OJEC) procurement procedures were used.

Devon were advised to take the 'restricted' route – which limits negotiation, but can be considered quicker. This works well where the outputs required are clearly defined, but not if you wish to explore options with potential bidders – which, with hindsight would have been the preferred option for the maintenance agreement.

The procurement process for the framework partnerships fell into three distinct phases:

Expression of interest (in response to the OJEC notice); to which there were 88 responses in which bidders were required to demonstrate minimum levels of competence and adequate financial standing, as well as relevant experience in a partnering environment. Areas of specialism such as Health and Safety were assessed by specialists in that field.

Tender (the written proposal); 15 bidders were selected to tender against a document which set out the aspirations for the County Council with particular emphasis on developing a collaborative relationship with the selected partners. A broad cross-section of Council staff was used to evaluate the sixteen sections of the bid with 70% of the marks being allocated to the quality of those sections. 30% of the marks were allocated to the Cost Section which included figures bid for profit and overheads, and an assessment of the projected capital cost savings to be made over the life of the framework. Every step of the process was monitored by consultants Mouchels, to provide a robust audit trail.

Tender (site visits and interviews); short-listed bidders were visited by a small procurement team to verify the content of the bid document, and to ask searching questions about partnering. This also allowed the team to get a 'feel' for the culture of the organisation, and the maturity of it partnering ethos. Interviews were conducted with bidders and exercises set to test their team working – the latter was of particular value in assessing the key team being put forward to run projects.

The final outcome was the selection of 6 partners to deliver major capital schemes in a programme worth up to £50million and 4 partners to deliver maintenance and minor works where the annual value is £9million.

Key points:

- The process is highly resource intensive, and requires a team to be dedicated full-time.
- Commitment is required at the highest levels within the organisation including the political will.
- Legal and audit departments should be fully engaged in the process.
- Research as widely as possibly – there is a surprising amount of goodwill to be drawn on.
- Make use of consultants for key areas of expertise.
- Don't underestimate the knowledge that resides within your own organisation.

- Consult as widely as possible – particularly important to ensure buy-in from staff and clients

Source: Chris Jackson, Partnering and Framework Manager, Devon Property, Devon County Council

Another example with more ambitious objectives and greater scope is provided by Defence Estates, an agency of the UK's Ministry of Defence, who are adopting Prime Contracting which includes many of the key elements of partnering, TQM and SCM. Its aim is to promote collaboration through leadership, facilitation, training and incentives, and replace short-term, contractually-driven, project-by-project, adversarial relationships with long-term, multiple-project relationships based on trust and cooperation. It includes the restructuring and integration of project processes and supply networks with fewer strategic supplier partners. These new relationships incorporate continuous improvement targets to reduce costs, enhance quality, and focus on the whole-life cost and functional performance of buildings (Holti *et al.*, 1999).

Some clients and their advisers are adopting the tiering structure outlined earlier in this chapter. In this approach it is the responsibility of the customer tier to organise, communicate and nurture the level below. Thus, the assembler takes responsibility for the welfare of the first tier suppliers, the first tier for that of the second tier firms, and so on down the pyramid as shown in Fig. 7.6. Given the leading role played by clients, they are at the apex of the pyramid with cost and design con-

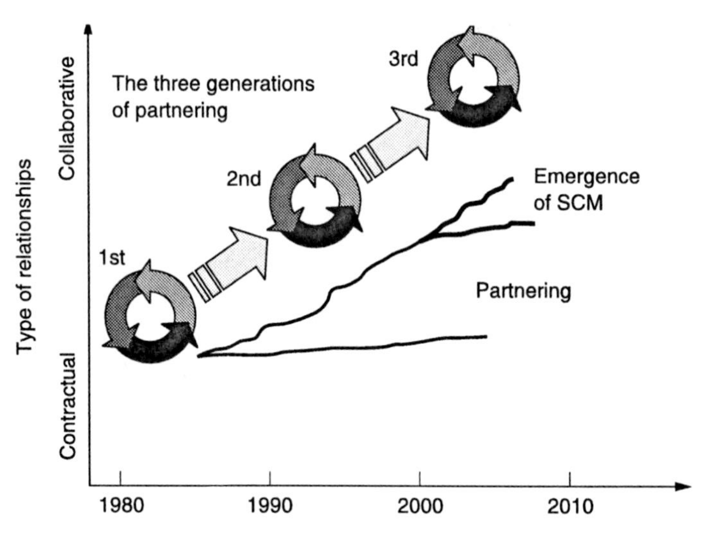

Fig. 7.13. Transition from strategic partnering to SCM

sultants and the main contractor as their first tier suppliers. The second tier comprises the specialist and trade subcontractors, with their material and component suppliers forming the third tier.

A variation on this structure is where the main contractor occupies the apex of the pyramid. This is rather like the Prime Contracting approach and cluster structure being adopted by Defence Estates. The cluster structure, which is illustrated in Fig. 7.14, is an idea developed by the Reading Construction Forum and piloted by Defence Estates in the 'Building Down Barriers' project. In this approach, a Prime Contractor is appointed to work with key supply partners, known as Cluster Leaders, who set the general direction of designing and delivering a significant element of the building – the groundworks and substructure, the superstructure, the services, and so on.

This allows the Prime Contractor to improve in an integrated fashion the process for designing and delivering the overall building as well as the materials and the components that go into its main elements. It means the client has a single contractual relationship with the construction team. The contract is held by the Prime Contractor, who is usually a main contractor, although the role can be undertaken by other members of the construction team. The team is stable and consists of designers, the prime contractor and suppliers who have a long-term contract with the client for a series of projects. This means that the team's performance can be improved from project to project and communication across the interface structures developed as shown in Fig. 7.15.

There is growing evidence of the effectiveness of SCM in certain parts of the industry. It appears to be most effective in the case of fairly standard buildings for regular and frequent clients of the industry, as shown in the ASDA case study.

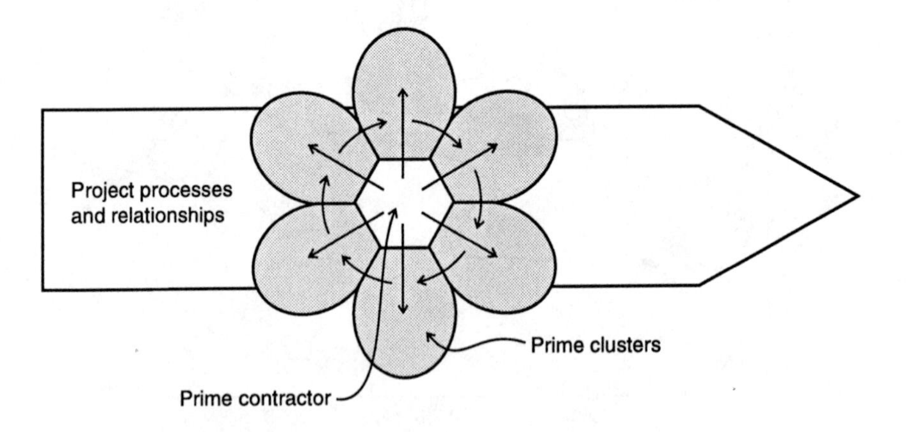

Fig. 7.14. Structuring of supply system using clusters

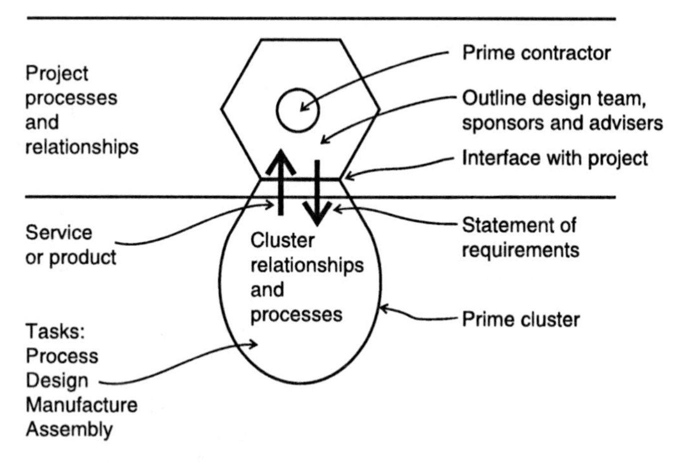

Fig. 7.15. Interface between the project and a cluster

Case study 7.5: ASDA – benefits of SCM in construction

Contractor Kajima UK Engineering completed a 9250 m^2, £11.5 million Asda superstore in Swansea under a *JCT81* contract. The project is based on Asda's model store, which is a standard design modified to suit each site. The building has a steel frame structure with a standing-seam built-up roof, composite cladding and brickwork panel walls. The floor layout is based around a central open-plan sales area. A two-storey office and restaurant block is located in front of the sales area, with a warehouse and plant area at the rear.

Key features

The Swansea project made a cost saving of 5.75% on a similar Asda store built in Gateshead 18 months earlier. This gives the store a benchmark score of 65% on the Government's key performance indicator for construction costs, which means it is 15% better than the industry average (50%) for all projects.

- The store was completed in 15 weeks, 39% faster than the Gateshead store. This gives it an 85% score on the indicator for construction time – 35% better than the industry average.
- Kajima used modular units to speed up construction time.
- Efficiency was improved by performance monitoring. Site processes were analysed to assess how much time was spent on productive and non-productive activities.
- Asda had post-project workshops with design team members and subcontractors to discuss what should be carried forward from this project.

The contribution of modularisation to cutting build time

'Asda was looking to make a quantum leap in the construction process,' says Richard Nicholls, the retailer's store development manager. 'The project aim was to bring down construction time without compromising safety or increasing costs, while producing a replicable build process that could be carried forward to the next project.' Asda has a standard store model that is adapted to suit each site. It also uses a set of partnering contractors and consultants to achieve year-on-year savings of 20% for store construction time and 10% in store costs. But to facilitate its 'quantum leap', Asda appointed Kajima, a contractor not included on its partnering list. The lessons learned from previous projects were made available to Kajima. The store was constructed in 15 weeks, following the completion of enabling works. This compared with the 26 weeks taken to construct a similar sized store in Gateshead 18 months earlier.

Cutting construction time means that more activities have to be carried out at the same time. Kajima proposed several changes to Asda's standard construction model, including using modularisation to speed up the build time and to take activities off the critical path. At the front of the store, on either side of the main entrance, are the double-storey office and restaurant. Kajima modularised these by bringing them to site as 28 fully fitted-out modules, taking them off the critical path. However, installing the units proved to be more difficult than the team had anticipated. 'It was an experiment,' says Neil Sargent, Kajima's contracts manager. 'Ultimately, if you are looking to bring down construction time, you have to start taking work off site. The lessons learned from this will be applied to future projects.'

Modular construction was also used in other parts of the building. Kajima clad the first 3 m of the external facade using modularised, brick-faced panels rather than building brick walls in situ. 'The wall was constructed in February,' says Sargent. 'By using wall panels we took weather dependency out of the operation.' The panels also meant the mess associated with a wet trade such as bricklaying was avoided, and there was no need to have scaffolding around the outside of the building, freeing up space on site. Kajima also used modularisation for the mechanical, refrigeration and electrical plantrooms. These were supplied as packaged units, fully assembled and commissioned, and installed in a separate area at the rear of the building. By mounting the units separate to the main store, Kajima ensured that they, too, remained clear of the critical path.

How monitoring improved productivity

One area targeted for improvement by Asda was site productivity. Even before Kajima had been awarded the contract, it arranged a three-day workshop with its selected subcontractors and Asda's team. This was to develop the process through which the project objectives would be achieved. These objectives included building the store in less than 20 weeks, and for 10% less than a similar store built a year earlier. Before starting on site, Kajima gave all its subcontractors an induction to explain how the teamwork culture would work and to allow them to commit to delivering the objectives. Sargent describes this as 'creating the atmosphere to allow operatives to challenge the obvious, to try to think outside their particular box and to think of others'.

Asda used Calibre, the Building Research Establishment's site performance monitoring system, to measure who was doing what, how much time they were taking, how much their activity contributed to advancing the project and how much time was spent on non-productive activities. 'We wanted to know where we could do better,' says Mike Abel, Asda's general manager of research and development. All site operatives were given an identifying number to wear on their vest, allowing the two BRE observers to record who was doing what task, and the time it took to do them. To ensure communication between the trades, Kajima insisted on meeting in a 'huddle' every morning and evening. The idea was borrowed from Asda's management team, and was seen as a chance to sort out problems.

At Swansea, the results of the previous day's Calibre monitoring were reviewed by the contractors and subcontractors at the morning huddle. The aim was not to look at what went wrong but at how the process could be improved the next day. For example, Calibre showed that the productivity of the lining contractors had decreased. At the huddle, the problem was found to have been caused by ductwork stored on site, which was restricting the liners' access. Calibre showed that the Swansea team spent 8% more time on productive activities than the industry maximum from previously monitored sites. The lower figures at the start of the project were caused by the high number of managers on site relative to the number of operatives. Neil Sargent, Kajima's contracts manager, thinks the high productivity figure can be attributed to a change in the attitude of everyone working on the project – something that was targeted before the project was on site.

Andy Pearson, The Benchmark series, *Building*, 9th July, 1999

The results from the survey undertaken by the authors, as shown in Fig. 7.16, also provide evidence of the potential benefits including greater value for the client, better integration between design and construction, enhancing quality and improving relationships and trust. This confirms that construction organisations have high expectations of SCM in addressing the three main problems facing construction:

- adversarial relationships
- fragmented processes and
- lack of customer focus.

The Cannons and Mace case study provides a further demonstration of the benefits to be derived from an effective implementation of SCM over time. It also highlights some of the difficulties related to the underperformance of a few suppliers and the need to increase their capabilities.

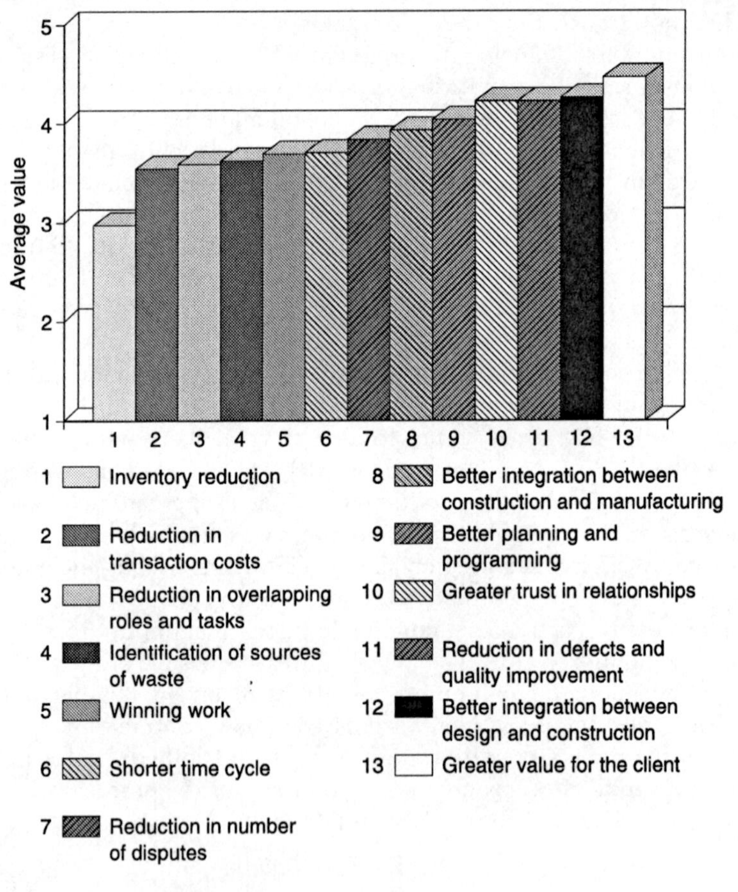

Fig. 7.16. The benefits of SCM in construction

Case study 7.6: Cannons and Mace – benefits to be derived from an effective implementation of SCM

Leisure operator Cannons unveiled an ambitious expansion programme in 1997. The aim was to treble the company's health and fitness clubs in three years, from 16 to 44. If the programme was to be successful, however, Cannons knew that the way it procured construction services would have to be changed. Over the previous three years, the operator had experienced all kinds of problems with contractors. 'To budget, on time and at quality appeared to be optional extras,' says Ken Redman, Cannons' technical director. The company's policy of using fixed-price competitive tendering had inevitably led to claims.

The company leases purpose-built shells which it then fits out as health and fitness clubs. But it felt that the construction problems it had experienced over the past three years had distracted from its core business as a leisure centre operator. 'Cannons is a leisure operator, not a property developer,' says Redman. Cannons also realised that its construction process was wasteful and inefficient. Although all the health clubs had similar features this was not recognised in the construction process, with each project being treated individually. 'We were reinventing the wheel every time and there was no security of delivery,' says Redman.

This was where Mace came into play. In 1997, Cannons brought the specialist on board for its construction management skills. However, Cannons quickly found that using construction management did not solve all of its problems as it still had the distraction of dealing with the subcontractors and it had to shoulder most of the risk for each project. But by keeping the same construction team together for several projects, Cannons found that it was able to move from construction management to an informal partnering arrangement with the core team of employer's agent, architect, and M&E design-and-build contractor.

As the partnering arrangement developed, Mace's role changed from that of construction manager to main contractor. And because the volume of work was so high, Cannons brought on board two other contractors – Pearce Leisure and Dutch contractor Pellikann – to add 'commercial tension' and to ensure that it was getting value for money. This partnering arrangement enabled Cannons to move to a fixed-price design-and-build contract on the basis that risks to the contractor would be minimal. And, crucially for Cannons, the move meant that it was now producing a generic design with which it became familiar, rather than repeating the process each time.

To differentiate itself from the other contractors, Mace decided to improve price and value by streamlining the fit-out process. The first act was to remove repetition from the initial stages by developing a generic brief for design and specification. This included production of a common internal layout for the clubs, along with standardised construction details for areas such as the changing rooms and crèche. 'It allowed us to introduce standardisation to the roll-out,' says Mark Reynolds, a partner at Mace. At the same time, the contractor worked with its partners to calculate what a scheme should cost per square foot. This information provided a useful way of comparing the costs of a scheme. A more detailed cost model was then produced to enable quick feasibility studies to be undertaken for a given development proposal.

To get a clearer picture of the construction process, Mace mapped the supply chain. This offered a better understanding of how the supply chain was configured and increased the predictability of delivery: if any hold-ups occurred, it could locate which firm was responsible and take steps to solve the problem or find an alternative supplier. The map could also be used to check whether the materials being sourced were good value for money. One problem that the mapping highlighted was with tiles for the changing rooms. These were on the project's critical path, but were frequently delivered late. Mace therefore proposed a change: tiles would be trucked to a central store before they were required, ensuring that supplies were always on hand.

On site, Mace introduced a productivity management system, known as Last Planner. This was to 'ensure that work plans were assigned each week', says Reynolds. This, in turn, helped managers to anticipate problems, such as a lack of men or a shortage of materials, and it encouraged communication. 'It commits all key contractors to saying what they are doing that week. If it is not completed, it allows us to ask why,' explains Reynolds. As the roll-out programme continued, Cannons sent the architects back to the clubs to ask the operators how well the centres were performing. Feedback from these revisits was then used to revise the brief.

Now, three years into the approach, Cannons has drawn up formal partnership arrangements with more than 15 companies and partnering is being embraced in earnest. 'We were not interested in the lowest priced contractors; what was important to us as operators was the ability to deliver,' says technical director Redman. 'The supply chain brings added value to the project – we don't talk about budget, time and quality any more. We have a group of people who understand our business operation.'

Reynolds agrees: 'By using the same team for each project, they develop an understanding with each other,' he says. 'The knowledge is in the supply chain to enable quicker programmes, better quality and added value by forcing down the cost in other ways,' explains Rachel Vincent, project manager at Mace. 'We're now at the stage where key suppliers are on board from the start in design development meetings.'

After three years of partnering with the same consultants and contractors, Cannons has been able to redefine the whole building process. Initially, there were eight main project stages from inception to completion. However, the stage for selecting contractors is no longer necessary and the employer's requirement document is also redundant. That document defined what is required in a health club – 'but we know that', says Reynolds.

Now there are only four stages to the project and the entire process, from initial briefings to completion, has been cut from 52 weeks to 28. The client undertakes the feasibility study and an outline design to make sure the scheme is deliverable. Then the whole team is brought on board. 'It's an open-book arrangement with the contractor – this is the budget, there is the team, you know the price, now go away and do it,' says Paul Taylor, production leader at Cannons.

A more recent innovation by Mace, created to speed information flow, is a project extranet. This enables the team to collaborate and share information from any location. All drawings, specifications and generic information are contained on the site, as is costing information. And once the project is finished, the documentation forms the basis of Cannons' operation and maintenance manuals.

'In the early days, we had a very traditional relationship with the specialist contractors,' says Reynolds. However, as relationships have developed, the key specialist contractors are brought on board at the outset. Their input is invaluable to ensure buildability and add value to the project. For example, it was a subcontractor that pointed out that if the plantroom was accessible from the outside of the building, it would be more convenient for the contractor during construction and for maintenance staff once the building was completed. Typically, 80% of the suppliers on a project are from the previous team, so they already know what needs to be done. This paid dividends at the Warwick centre, by allowing Mace to ensure that the gym was fitted out much earlier than usual to allow Cannons' sales team to set up shop and sell membership in advance of the centre's opening. Usually, the sales team would have to lease and fit-out a series of temporary sales cabins. 'It saved our operations quite a bit of money,' says Redman.

The biggest problem faced by the contractors has been the poor performance of a few suppliers, some of which have been dropped. The group causing the most difficulty, says Reynolds, is painters and decorators, because they tend to be regionally based. The next step for Mace will be to assist the best suppliers and help them develop their supply chain. 'Since we've been partnering, every one of our new sites has been delivered ahead of schedule,' says Redman. 'In two years we've brought our construction time down from 24 weeks to 16, costs have been static and our partners are giving us more for the same money.'

Summary of key benefits:

- 11% reduction in capital cost, while simultaneously increasing the quality of the product
- 25% reduction in construction time
- approaching 100% predictability
- approaching zero defects at handover
- zero reportable accidents throughout the entire programme.

Movement for Innovation, Case History, www.m4i.org.uk

There are considerable difficulties in applying SCM in construction as shown in Fig. 7.17. These include short termism, lack of trust and adversarial relationships, the transient nature of construction projects and the significant number of irregular clients. Revealingly, a 'lack of internal preparedness' was not seen as a serious obstacle.

Given its traditional short-term, contractual relationships and fragmentation, adopting a more collaborative approach is not straightforward. The integration of organisations with different cultures, power and knowledge bases can be problematic in construction given its multiple and often hidden goals; power imbalances; differences in professional language, culture and procedures; incompatible collaborative capability; and the tension between autonomy and accountability. Addressing these differences provides the basis of the benefits to be derived from SCM, but they also explain the reluctance and slowness of many construction organisations to meet the challenges and complexities of working more closely together. The main barriers which need to be addressed in implementing SCM in construction are outlined in Table 7.3.

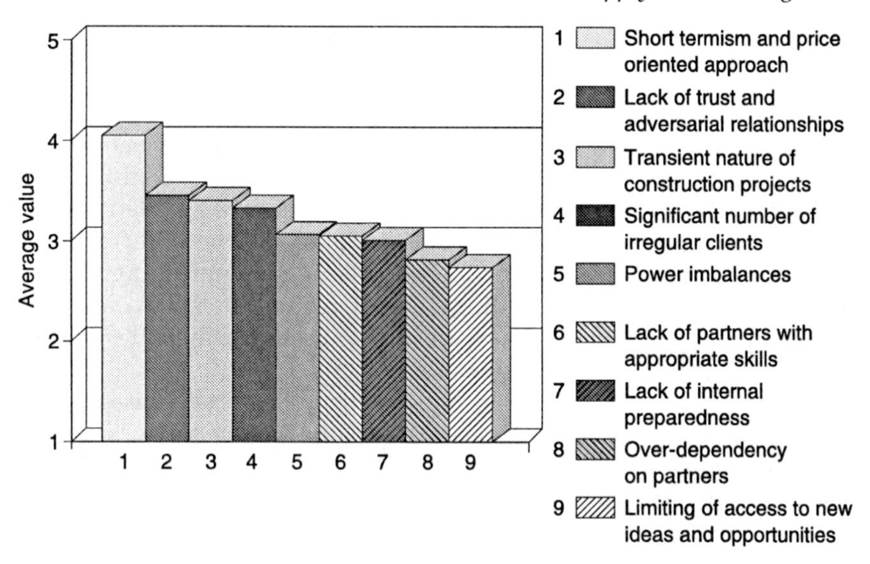

Fig. 7.17. The main problems associated with SCM in construction

Table 7.3. The main barriers to the implementation of SCM in construction

- 80% of the industry's clients with their small or infrequent construction programmes – this means that only 20% of construction's clients have the commitment, knowledge and the necessary leverage to adopt network competition.
- The transient and short-term nature of construction projects, processes, teams and relationships.
- Present procurement strategies with their emphasis on contracts and price competition.
- Deeply embedded adversarial relationships and opportunistic behaviour.
- Fragmented processes.
- Lack of possible partners with the appropriate collaborative capability.
- Multiple and hidden goals.
- Major power imbalances.
- Lack of contractual and competence trust.
- Insufficient resources and time to build relationships, integrate processes and manage logistics within a one-off project environment.
- Differences in professional language, culture and procedures.
- Lack of experience of innovations such as JIT and TQM.

Conclusion

Effective SCM requires stable and long-term relationships between organisations in supply chains. Clearly, this basic requirement exists in the case of the regular construction client such as Sainsbury's, ASDA WAL*MART and Defence Estates where effective long-term relationships have been developed. In the case of significant and frequent customers of construction services, elements of SCM including closer long-term relationships are leading to greater transparency in transactions, trust and commitment which are seen as being central to the development of competitive advantage.

However, this is very difficult for infrequent clients of construction and their one-off projects. Such clients have little opportunity or indeed motivation to stabilise and improve their construction suppliers. Centrally coordinating and integrating their process with project processes is difficult where the client is unwilling, or unable, to exercise this degree of leadership and coordination. In these circumstances, leadership and management of supply chains will need to be undertaken by other project participants such as main contractors. However, these barriers to implementing SCM in construction also indicate that it offers a highly relevant approach to improving its performance.

If appropriately implemented, SCM can offer a way forward for improving relationships, integrating processes and increasing customer focus. The benefits to all participants can include, improvements in quality and delivery, more repeat work, reductions in the overheads associated with obtaining work, increasing profitability, and the acquisition of new specialist knowledge and skills in significant sectors of the construction market.

SCM can help construction manage the relationships and processes between the different participants by providing a more inclusive environment where construction organisations such as specialist and trade subcontractors and suppliers can be given a more participative role in order to ensure greater coordination of an increasing part of the process. Both, the upstream processes to the client and the end-user, and downstream processes involving the specialist and trade subcontractors and their suppliers, need to be more effectively and fully involved as shown in the BAA case study.

Case study 7.7: BAA – customers and suppliers reap the benefits of long-term relationships

BAA spends about £35 million p.a. on the pavements at Heathrow, Gatwick, Stansted and Southampton. In the mid-1990s, they invited AMEC Civil Engineering to work in partnership to procure £130 million of projects over five years. AMEC were keen to improve cash flow and break the cycle of unpredictable profits and losses. BAA believed that cost savings in the region of 30% over the life of the framework were obtainable.

A five-year Framework Agreement led to back-to-back frameworks, strategic supplier groups, supplier clubs, mentors and benchmarking. The management structure utilises BAA's client skills in project and commercial management, while exploiting AMEC's contractor expertise in planning and operations. Responsibility for development was to be shared. A jointly agreed general manager who is a BAA employee leads the team. Within the team, some BAA personnel are directed by AMEC managers and vice versa.

The BAA Resident Engineer's team were initially sceptical, fearing their jobs were at risk. It took some time to move them into QA roles within the construction team. A few people in BAA and AMEC found it difficult to let go of the adversarial culture but most were won over by early successes. Those unable or unwilling to adapt were transferred to other projects.

With 50 projects completed, the Pavement Team has moved a long way from traditional shadowing of roles in the early days to a fully integrated culture. Management appointments are now made on ability, irrespective of company affiliation. Integration has increased trust and that promotes further integration. Recently, second-tier suppliers have accepted management roles in the team, strengthening further the concept of partnering the supply chain. The management team is continually adapting to further strengthen the supply chain. Having successfully integrated second-tier suppliers into the Pavement Team, negotiations to draw in third-tier suppliers are under way.

Framework agreements are ideally suited to long-term client-contractor relationships for the operation and management of infrastructure assets. The integrated team structure is also effective where project teams work together to deliver a series of one-off projects of a similar generic type.

Movement for Innovation, Case History, www.m4i.org.uk

Extending a comprehensive SCM approach to the whole of the construction market will be difficult and challenging. A greater part of the industry needs to be more adept at segmenting the market, identifying and developing critical supply chain assets, being aware of market conditions, and aligning all project and supply chain processes to meet the needs of the external customer.

There is strong evidence that there is a growing awareness of the concept of SCM among construction practitioners. However, it must be recognised that it is too early to undertake a comprehensive evaluation of its implementation and its impact on construction. Also, it is clear that the present complexity, fragmentation, interdependency and uncertainty which characterise the construction market and process will influence the way in which SCM and other innovations are adopted and implemented.

In addition, construction is moving towards the adoption of SCM as a fifth generation innovation without having benefited from earlier innovations, such as JIT and TQM. It is only relatively recently, with the emergence of partnering, that the industry has started moving towards more collaborative relationships and integrated processes that can be interpreted as laying the foundations for SCM. Even where partnering has been adopted, it is still largely misunderstood and has not yet led to a widespread change of culture, which remains essentially adversarial with arm's-length relationships and a significant use of price-competitive procurement approaches and rigid contracts. In addition, partnering is mainly being adopted upstream and essentially between clients, consultants and main contractors and has yet to be extended to many of the suppliers downstream of the main contractor.

There are significant difficulties which need to be addressed if SCM as a fifth generation innovation is to be effectively implemented. These include:

- the lack of preparedness of construction organisations to adopt SCM
- the limited understanding of the concept and the prerequisites associated with its implementation
- the unwillingness to rationalise supplier and customer bases
- the difficulty in establishing a clear common purpose, exchanging information and sharing learning.

This can be interpreted as a further indication of the industry's lack of awareness and reluctance to embrace a new culture associated with SCM relationships. Although learning is perceived as increasingly important in supporting innovation in construction, the types of learning being undertaken do not match the competencies and the cultural changes needed for such a complex, multi-factor and dynamic innovation. A further indication of the limited understanding of the scope and complexity of this type of innovation is demonstrated by the low importance attached by construction to external support. The current external support does not match the complexity and challenges associated with the fifth generation innovation.

References

AKINTOYE, A., SMITH, G. and SAKU, J. A survey of supply chain collaboration and management in the UK construction industry. Supply Chain Management in Construction – Special Issue, *European Journal of Purchasing & Supply Management*, **6**, 2000, 159–168.

ALI, F. *et al.* Developing buyer-supplier relationships in the automobile industry, a study of Jaguar and Nippondenson. *European Journal of Purchasing and Supply Management*, **3**, No. 1, 1997, 33–42.

AXELROD, R. *The Evolution of Cooperation*. Penguin, London, 1984.

BALL, M. *Rebuilding Construction: Economic Change in the British Construction Industry*. Routledge, London, 1988.

BURGESS, R. Avoiding supply chain management failure: lessons from business process re-engineering. *International Journal of Logistics Management*, **9**, 1998, 15–23.

CHRISTOPHER, M. Logistics and Supply Chain Management – Strategies for Reducing Cost and Improving Service. Financial Times – Prentice Hall, 1998.

CHRISTOPHER, M. and JÜTTNER, U. Developing strategic partnerships in the supply chain: a practitioner perspective. *European Journal of Purchasing and Supply Management*, **6**, 2000, 117–127.

COOK, J. and WALL, T. New work attitude measures of trust, organizational commitment and personal need non-fulfillment. *Journal of Occupational Psychology*, 1980, **53**, 39–50.

COX, A. and TOWNSEND, M. *Strategic Procurement in Construction: Towards Better Practice in the Management of Construction Supply Chains*. Thomas Telford, London, 1998.

DASGPUTA, P. Trust as a commodity in Gambetta, D. (ed), *Trust, making and breaking and cooperative relations*. London, Sage, 1988.

DUBOIS, A. and GADDE, L. Supply strategy and network effects – purchasing behaviour in the construction industry. Supply Chain Management in Construction – Special Issue, *European Journal of Purchasing & Supply Management*, **6**, 2000, 207–215.

EDUM-FOTWE, F. T., THORPE, A. and MCCAFFER, R. Information procurement practices of key actors in construction supply chains. *European Journal of Purchasing and Supply Management*, **7**, 2001, 155–164.

FARRIS, G. F. Trust, Culture and Organizational Behavior. *Industrial Relations*, **12**(2), May 1973, 144–157.

GOOD, D. Individuals, interpersonal relations and trust. In GAMBETTA, D. (ed.). *Trust: Making and Breaking Cooperative Relations*. Blackwell Publishing, Oxford, 1988, pp. 31–48.

HARLAND, C. M. *Supply Chain Management; Relations, Chains and Networks*, 1996.

HARLAND, C. M. *et al.* Developing the concept of supply strategy. *International Journal of Operations and Production Management*, **19**, No. 7, 1999, 650–673.

HILL, T. *Operations Management – Strategic Context and Managerial Analysis*. Macmillan Business, London, 2000.

HINES, P. *Creating World Class Suppliers*. Pitman Publishing, London, 1994.

HOFSTEDE, G. *Cultures' Consequences, International differences in work related values*. Newbury Park, London, Sage, 1980.

HOLTI, R. *Adapting supply chain for construction*. Workshop Report, CPN727, Construction Productivity Network, CIRIA, 1997.

HOLTI, R., NICOLINI, D. and SMALLEY, M. *Prime Contracting Handbook of Supply Chain Management Sections 1 and 2*. Tavistock Institute, London, 1999.

HUXHAM, C. (ed.) *Creating Collaborative Advantage*. Sage Publications, London, 1996.

JAMES, P. C. and SAAD, M. The role of trust in the creation of supplier associations: the example of a Welsh partnership. Paper given at the *Fifth International Conference on Multi-Organisational Partnerships & Co-operative Strategy*. 6, 7, 8 July 1998, Balliol College, Oxford.

JONES, M. and SAAD, M. *Unlocking Specialist Potential: A More Participative Role for Specialist Contractors*. Thomas Telford, London, 1998.

KANTER, R. M. and MYERS, P. Inter-organizational bonds and intra-organizational behavior: how alliances and partnerships change the organizations forming them. Paper presented at the *First Annual Meeting of the Society for the Advancement of Socio-Economics*, Cambridge, MA, 1989.

KELLERMAN, K. 'Information exchange in social interaction' in interpersonal processes: *New directions in communications research*. Roloff, M. E. and Miller, G. R. (eds). London, Sage, 1987.

KOSELA, L. Management of production construction: a theoretical view. *Proceedings of the Seventh Annual Conference of the International Group for Lean Construction IGLC-7*, Berkeley, July 26–28, 1999, 241–252.

KREPS, D. M. Corporate culture and economic theory. In SHEPSLE, J. E. A. K. A. (ed.), *Perspectives on Positive Political Economy*. Cambridge University, New York, 1990, pp. 90–143.

LAMMING, R. *Beyond Partnership: Strategies for Innovation and Lean Supply*. Prentice Hall, New York, 1993.

LIPPARINI, A. and SOBRERO, M. The glue and the pieces: entrepreneurship and innovation in small firms' networks. *Journal of Business Venturing*, **9**, 1994, 125–140.

MOWERY, David C. (ed.). *International Collaborative Ventures in US Manufacturing*. Ballinger, Cambridge, MA, 1988.

NAYLOR, J. *Introduction to Operations Management*. 2nd edn. Financial Times – Prentice Hall, 2002.

NEW, S. and RAMSAY, J. A critical appraisal of aspects of the lean approach. *European Journal of Purchasing and Supply Management*, **3**, No. 2, 1997, 93–102.

ROTTER, J. B. A new scale for the measurement of interpersonal trust. *Journal of Personality*, **35**, No. 4, 1967, 651–65.

SAAD, M., JONES, M. and JAMES, P. A review of the progress towards the adoption of supply chain management (SCM) relationships in construction. *European Journal of Purchasing & Supply Management*, **8**, 2002, 173–183.

SAKO, M. *Prices, quality and trust: inter-firm relations in Britain and Japan.* Cambridge University Press, Cambridge, 1992.

SAKO, M., LAMMING, R. C. and HELPER, S. M. Good news – bad news. *European Journal of Purchasing and Supply Management*, **1**, No. 4, 1994, 237–248.

SLACK, N. *et al. Operations Management*, 3rd edn. Financial Times – Prentice Hall, 2001.

SMELTZER, L. R. The meaning and origin of trust in buyer-supplier relationships. *The International Journal of Purchasing and Materials Management.* Winter, 1997, pp. 40–48.

SPEKMAN, R. E., KAMAUFF Jr, J. W. and MIHR, U. An empirical investigation into supply chain management: a perspective on partnerships. *International Journal of Physical Distribution and Logistics Management*, **28**(8), 1998, 630–650.

TAN, K. C. A framework of supply chain management literature. *European Journal of Purchasing and Supply Management*, **7**, 2000, 39–48.

TOWIL, D. R. Time compression and supply chain management – a guided tour. *Supply Chain Management*, **1**, No. 1, 1996, 15–27.

VOLLMAN, T. *et al.* Supply chain management. In *Mastering Management.* FT Pitman, London, 1997, pp 316–322.

WHIPPLE, J. M. and FRANKEL, R. Strategic alliance success factors. *Journal of Supply Chain Management*, **36**, No. 3, 2000, 21–28.

WOMACK, J. P. and JONES, D. T. *Lean Thinking: Banish Waste and Create Wealth in your Corporation.* Simon and Schuster, New York, 1996.

Chapter 8

Key factors conducive to innovation in construction

The main objective of this chapter is to reflect on the main inhibitors to innovation in construction and to propose areas to be addressed in order to manage and diffuse innovation more effectively within the industry.

A number of construction organisations are continuing to innovate in relation to their products, processes and operating systems. They are also continuing to respond to changes from their external environment including their markets and regulatory and institutional framework. However, as discussed throughout this book, there are still substantial inhibitors to the adoption of post-Fordism approaches in much of construction. The implementation of innovation remains impeded, as demonstrated in Chapters 3 and 4, by its limited response to the following key factors:

- awareness of the need to change
- responsiveness to internal and external change
- linkages within and between organisations
- strategic, holistic and systematic approach
- culture conducive to learning and innovation
- commitment.

In addition, the recent substantial increase in the level of external support is still not reaching the majority of construction organisations. Also the nature of this support is not yet appropriately reflecting the features of innovation associated with post-Fordism.

Inhibitors to innovation in construction

There are considerable barriers to innovation inherent in construction's specificities and culture, which have been identified by this book. These barriers include a lack of appropriate leadership, insufficient learning, lack of investment in people, inappropriate organisational structures and management approaches which remain influenced by a strong Fordist orientation. There is a continuing misalignment and fragmentation of upstream and downstream processes, adversarial relationships, emphasis on price competition and the exclusion of many key participants such as irregular clients and specialist and trade subcontractors and their suppliers. The continuing emphasis on price competition is, for instance, pushing main contractors to place greater focus on the upstream side of the process while largely under-utilising the potential in the downstream side. There is also some evidence that efforts to diffuse innovation downstream to other participants such as specialist and trade subcontractors is being inhibited by some participants concerned about losing their influence within traditional processes and relationships.

The current types of learning are not appropriate and not sufficient to generate the shared learning required for an effective implementation of fifth generation innovation. The fifth generation innovation is associated with greater integration, the use of networks and a strong emphasis on on-going learning. Most training to support implementation is 'in-house' or one-off day courses and events. They are essentially driven by a short-term perspective, with an undue emphasis on technical rather than people issues. Training is also still influenced by a strong functional approach in both project and organisation and hence compartmentalised and lacking a holistic and systemic approach. There is insufficient emphasis on people and teamworking in construction in comparison with other industries. This is particularly prevalent in the case of specialist contractors. There is also a lack of awareness about the complexity of innovation and the external support required for the development of such a complex innovation is inadequate.

In spite of the many calls for changes in construction, the main champions of innovation remain essentially restricted to a small number of regular and significant clients, mainly in the private sector but increasingly in the public sector. This means that in the rest of the industry there is still insufficient interaction with the external environment despite the formation of the Construction Industry Board following the publication of the government-sponsored Latham report in 1994. However, this interaction with the external environment has, to some extent, intensified since the publication of the Egan Report in 1998 through, for example, the work of the Movement for Innovation (m4i) and the Construction Best Practice Programme. This pressure has continued with the publication of the report *Accelerating Change* in 2002.

Most of the above inhibitors can be addressed by ensuring better management and development of people, including:

- greater and more inclusive participation of specialist and trade subcontractors and their suppliers
- more effective leadership throughout the project process and the organisations involved
- appropriate and continuous learning at the individual, project and organisational level
- new culture to support, promote, implement and sustain innovation in construction and
- more effective external support.

A more participative role for specialist and trade subcontractors and their suppliers

Although there are some notable examples of downstream integration through a greater involvement of specialist subcontractors, construction has so far done little to give specialists a more essential role in its efforts to improve effectiveness and competitiveness through innovation. As discussed in Chapters 6 and 7, part of construction has started to adopt partnering and Supply Chain Management. However, most of this has been focused on upstream integration between clients, consultants and main contractors. Given their increasing significance within construction, specialist and trade subcontractors and their suppliers need to play a greater role in helping construction to develop and implement innovation. As they can account for as much as 80% of contract expenditure, this implies that improving the innovative capacity and overall performance and competitiveness of the construction process is strongly dependent on involving and motivating specialist and trade subcontractors and raising their performance as shown in the Arcadia case study.

Case study 8.1: Arcadia Group – early involvement of specialist subcontractors improves cost and delivery

In early 2000, Arcadia Group launched BrandMAX, the programme to re-engineer the company's property portfolio including Top Shop, Top Man, Miss Selfridge, Burtons, Dorothy Perkins, Wallis and Evans. Their mission was to refit 700 fashion retail stores in just nine months.

The success of BrandMAX builds on the lessons learned in fitting out Arcadia's new 28 000 ft^2 stores in the prestigious Southampton West Quay. This project was the turning point in procurement practice for Arcadia, moving from tendered to negotiated contracts. The big benefit has been extending the typical contractor mobilisation period from 3 to 14 weeks, thus involving specialist contractors early enough to make substantial budget and programme improvements. In BrandMAX, Arcadia set up rolling contracts with selected suppliers working in integrated teams, region by region, and they ran a shop-fit forum. The contractors were cautious initially, but by the end they were all talking to each other. It has evolved into a network of suppliers, and competitors are now supplying one another across regional boundaries.

BrandMAX has been a success, contributing to Arcadia's improved profits. Comparing the Southampton project with a similar but traditionally procured project at Milton Keynes, the reasons for this success include the following.

- Cost – early involvement of contractors empowered them to value-engineer solutions that reduced average unit costs by 8%. The extra lead time was used to programme the work so that areas where they knew the supply chain could make a big difference in design were delayed.
- Time predictability – shopfitting is possibly one of the most time-sensitive construction processes, because store opening dates are sacrosanct. The Southampton project was ready for occupation dead on time and defect-free after its twelve-week programme. But the Milton Keynes project was not fully desnagged until seven days after store opening. Prior to Southampton West Quay, about 5% of Arcadia's projects failed the 'open on time and defect-free' test but the failure rate has dwindled to just 0.5% in BrandMAX.
- Profitability – Arcadia's suppliers are generally finding that the new approach improves their business. Karen Friendship of shopfitting manufacturers Alderman Tooling says: 'It has certainly benefited our relationship and applying the *Rethinking Construction* principles is more profitable. But competition means it is difficult to hold margins up for long.'

The main lessons learned include:

- negotiating with selected contractors is producing better results than traditional tendering
- early contractor mobilisation gives the project breathing space to optimise the design
- specialist contractors and suppliers bring invaluable hands-on experience to the detailed design process
- regulating the workflow to trade contractor and suppliers improves the consistency of the workforce and minimises cost and disruption.

Movement for Innovation, Case History, www.m4i.org.uk

The importance of these subcontractors in the process is also emphasised in a number of reports such as the Latham report and the Reading Construction Forum's report *Unlocking Specialist Potential* (Jones and Saad, 1998) which argues their largely untapped potential to contribute to innovation.

In construction, individuals seldom work in isolation from others. Construction companies and projects are made up of interdependent groups of people. The working of groups and the influence they exert over their membership is an essential feature of the way people behave and perform their task. People and teams are important in the successful adoption of innovation. Evidence suggests that companies and projects function best when members are part of highly effective and cohesive work groups. The challenge is to transform construction groups into teams who learn and innovate together. However, interviews undertaken by the authors with participants in the construction process underlined the strong 'them and us' environment of mistrust and hostility in which consultants and contractors criticise the poor performance of the 'subbies', while the specialists feel that they are being undervalued by the consultants and contractors.

The most significant barriers preventing specialist and trade subcontractors from making a greater contribution to construction's performance are shown in Table 7.2. Many of these barriers and problems demonstrate the need for improvements in relationships through greater participation and empowerment. There is further potential for more innovation through similar and complementary improvements within the supply chains supporting the main clusters of specialist and trade subcontractors, as shown in Fig. 8.1.

Table 7.2 in Chapter 7, indicates that relationships between individuals and organisations in construction projects are often poor and create considerable conflict, impeding performance and innovation. Given these poor relationships, teamwork is seen as being fundamental to developing good intra- and inter-organisational relationships and the key to unlocking the potential of subcontractors and suppliers and improving their performance. It also facilitates the development of deeper people-to-people relationships. However, teambuilding can be difficult in the context of projects as it takes time and resources and it must not be rushed. Clearly, there is a need to allocate sufficient time for teams to reach the final and most productive stage in their development. Such investment in time is more likely to provide tangible returns in terms of motivation, commitment, learning and performance. However, current approaches to project management have largely failed to include time for such activity. The current trend to develop long-term relationships in construction and to extend the life of project teams through strategic partnering is presenting greater opportunities for more effective and appropriate

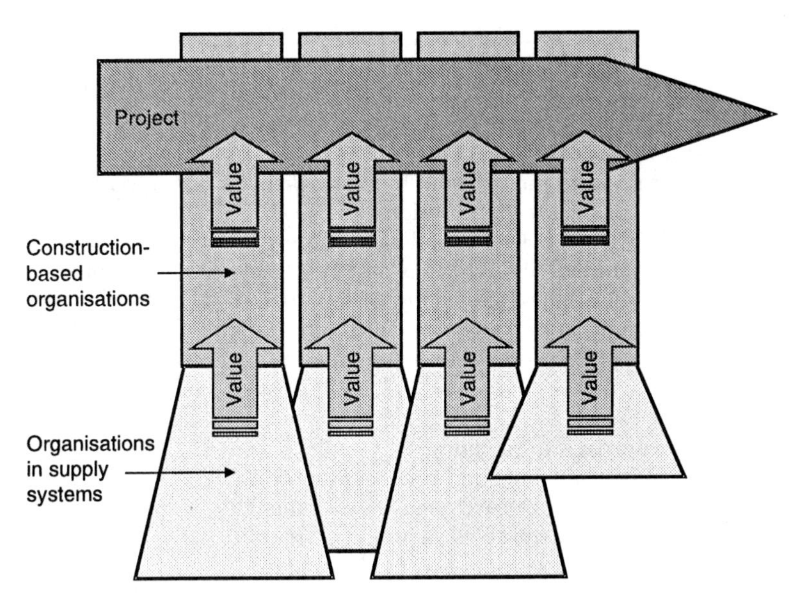

Fig. 8.1. Adding value through specialist and trade subcontractors and their supply chains

learning and other benefits to be acquired and passed to subsequent projects. This can lead to greater stability, trust and a climate more conducive to learning and innovation.

With the exception of some of the larger M&E organisations and the newly emerging sub- and superstructure specialist, many specialist and trade subcontractors are small firms and consequently have many strengths but also vulnerabilities which need to be recognised by the other participants in the process. Experience from the other sectors indicates that the other participants within the project and the supply chain must play their part in developing the strengths of subcontractors while at the same time helping them to address their vulnerabilities. Innovation is crucial to become and remain competitive (Trott, 2002). However, it is difficult for small enterprises to innovate on their own (Pietre van Dijk and Sandee, 2002). Innovation processes are uncertain, costly and very risky, especially in the case of SMEs which have weaknesses, as shown in Table 8.1, related to their lack of resources and skills which inhibit innovation. Innovation is also inhibited by their strategies in which they often compete through low prices, obtained by lowering their margins.

There is growing evidence that collaboration such as partnering between small enterprises, either in projects or in supply chains, makes individual small firms stronger. The idea is that commitment to learning and innovation takes place more easily in collaborative inter-organisa-

Table 8.1. Characteristics of innovation by SMEs

Strengths

- Specialist know-how and expertise
- Ability to respond to the needs of niche markets by taking on jobs that are too small for bigger organisations
- Exceptional drive and determination to succeed
- Ability to accommodate to changes and customer requirements
- Ability to respond quickly to changes
- A natural inclination to collaborate with other organisations in the process

Vulnerabilities

- Short-term approach to planning
- Under-investment in skills and management training
- Lack of investment in research and development
- Always perceiving themselves as junior partners
- Conditioned by a price-orientated environment
- Wariness of larger organisations
- Scepticism about sharing information and knowledge
- Most knowledge is acquired informally and is in a tacit form
- Tendency to be reliant on a small number of customers
- Problems in managing cash flow and raising money at acceptable terms for expansion
- Burden of complying with regulations and dealing with tax administration and paperwork

tional relationships as shown in the Ravenswood School project and Mowlem case studies.

Case study 8.2: Ravenswood School – supporting specialist and trade subcontractors

On the Ravenswood School project, main contractor Farrans trialled a process for engaging trade and specialist subcontractors and suppliers in the partnering team to use their knowledge of buildability early in the design. It was a learning process at Ravenswood, according to Suffolk's strategic implementation manager Len Taylor. 'We probably didn't develop the connections,' he admits. 'We found that most of the trade contractors and suppliers lacked knowledge and experience in school projects. This was particularly apparent in the mechanical and electrical work. We ended up retaining much of the design in the core team.'

But Suffolk has not been deterred by this experience and is developing their school product and intensifying trade contractor engagement in a follow-on project. Their timing looks right to take advantage of the £8.5 billion that the Government has committed to building, modernising and refurbishing schools over three years. Trade contractors and suppliers who are presently working in schools will become key players in the application of *Rethinking Construction* in this massive programme. Finding such firms with the right attitude is not easy, but it follows much the same principles that apply in selecting a main contractor. Partnering adviser Neil Carpenter of WS Atkins looks for evidence of what they really think. 'You can tell the difference between those going through the motions and those actually trying it,' he says.

Measuring progress is seen as important. In addition to the industry headline KPIs, the Ravenswood team measured some indicators of their success in engaging the supply chain.

- Design accuracy following the issue of construction drawings – the project manager labeled each additional drawing with a 'C' for a client change, 'D' for a design development or 'E' for a design error. The idea was to facilitate analysis and learn lessons about design accuracy.
- Package price versus final account – the difference between the price at the time of procurement and final account is an indicator of how well the work package was defined and priced. On average, the final package price was 4% more than the price at time of procurement, as budgeted.
- Time to settle package final accounts – the time taken from completion of the work package to settling the final account was a measure of how well the team collaborated. The average processing time was 11 weeks.

Overall, Suffolk's major successes were demonstrating how to successfully recruit a main contractor partner in just five weeks and retain cost control. The target cost was set at £917/m^2, some 11% lower than the rate calculated from the Building Cost Information Service data. The out-turn result of £900/m^2 was even better. The main lessons learned include:

- looking for suppliers with the leadership and vision to be team players
- specific project experience and design ability
- Recognise the importance of helping people to adapt to change

- devising incentives that address the supply for the clients big issues (e.g. retention) and the supply chain (e.g. incomplete construction and paperwork)
- encouraging good performers with further contracts that reinforce team working
- being disciplined and organised about learning from package debriefings
- removing the blame culture and ideas will flourish!

Movement for Innovation, Case History, www.m4i.org.uk

Closer inter-organisational relationships, characteristic of post-Fordism can help SMEs to raise their capacity to learn and innovate, taking advantage of the external economies and joint action induced by a common purpose based on mutual advantage. There is also a need for more systematic institutional support to help SMEs build up technological competence and innovative capabilities (Pieter van Dijk and Sandee, 2002).

Case study 8.3: Mowlem – supplier development

Feedback from Mowlem Building's customer surveys identified two areas for improvement – management of subcontractors and finishing on time. To increase the skill levels in the subcontractor companies it was decided to provide training in planning activities, enhance the effective management of resources and improve their ability to complete work on time.

A project was designed and piloted with funding from the European Social Fund. Participants from the subcontractors attend a two-day training course run by two external tutors. The course covers contract plans, programming techniques, scheduling, monitoring and controlling work, progress meetings, method statements, safety, communications and the cost of non-quality. Skill topics are covered through discussion, participation, brainstorming sessions, team exercises and a final role-play of a site progress meeting.

Richard Kochanski from Mowlem explains that 'Participants can question a member of their senior management team about issues and concerns they may have in dealing with us and other main contractors.' Following the training, three site visits are made by one of the tutors to the participants' place of work. The visits are to help and support participants in using the skills gained. A training-needs analysis is produced on the third visit to encourage life-long learning for the participant.

The outcomes of the project are as follows.

- Subcontractor supervisors are trained in site management and planning skills. These can promote improved workplace activities with better site quality, productivity, less wasted materials and costs. If their work is finished on time and reduced this has a direct benefit to Mowlem as the main contractor.
- Understanding that the pace of construction is important and to this has to be agreed by the team. Any increase or decrease in the pace can affect the trades before or after the activity.
- Increased site communications, such as progress meetings. This is an opportunity for pro-active problem solving and to remain aware of the ' big picture'.
- Better understanding of other people's problems and ways of doing your work to help others.
- Closer working relationships with selected subcontractors.
- Feedback from the subcontractors on the strengths and weaknesses of working with Mowlem – these are used by Mowlem for their own business improvement.

Mowlem Building

However, as mentioned in Chapters 6 and 7, the setting up of linkages requires commitment, sustained effort and special skills on the part of the organisations and the institutions involved to encourage such inter-organisational relationhips. In addition, those linkages are more difficult to set up when, for instance, as in construction, markets are fragmented, narrow and very often incapable of providing the environment needed to stimulate linkages. Forging inter-organisational relationships also has risks for small firms such as specialist and trade subcontractors, especially if the partner is a larger, stronger enterprise as in the case of many main contractors and clients. The small firm runs the risk of becoming over- dependent or, at worst, suffering exploitation at the hands of large organisations.

Leading and championing change

As discussed earlier, innovation is increasingly being recognised as a complex and challenging multi-factor process. Its implementation and management requires strong and effective leadership to bring about the substantial internal and external structural and attitudinal changes needed if it is to be successful. Leadership in this context comprises the processes by which individuals and organisations are empowered to work together in a closer and more collaborative way in order to support and promote innovation. This means that leaders and champions need to believe in and be totally committed to supporting learning and innovation.

Client-led innovation is becoming a significant force for change in some parts of construction as clients seek to drive down costs, reduce time and add value in their construction projects. By virtue of their position and influence in the development process, regular clients have a key role in spearheading change in the relationships between the main project participants and the way construction projects and processes are developed and managed.

Senior managers in client organisations have a pivotal role to play by leading by example and empowering others (within and outside their own organisation) to collective action by:

- visibly demonstrating their commitment to innovation and learning
- providing the appropriate front-end resourcing and assistance
- overcoming the resistance to cultural change
- developing a no-blame culture
- recognising and rewarding people's efforts and achievements.

In the case of a fifth generation innovation, this implies preparing people to do things differently and resourcefully and to fundamentally rethink business relationships, organizational structures, processes, roles and responsibilities. Meeting these challenges means abandoning much of the orthodoxy of conventional wisdom and current thinking and actively involving employees and partners in identifying issues and solving problems and creating new ways of working within an appropriate culture. With the development of the Construction Clients Forum, clients can benchmark their progress in leading this change in their supply chains and increasing customer focus. The Forum has developed guidelines to define the fundamental principles on which to build relationships within which customers and suppliers throughout the supply chain can work together to achieve and maintain performance improvements.

Case study 8.4: Construction Clients' Charter Framework

Through its Charter, the Construction Clients' Forum gives more specific guidance on 'leadership' and 'focus on the customer' for clients in general. This includes:

- providing client leadership, both for the improvement in procurement processes, and for the supply side to develop and innovate to meet clients' needs

- setting clearly defined and, where possible, quantified objectives and realistic targets for achieving these
- fostering trust throughout the supply chain by treating suppliers fairly and ensuring a fair payment regime
- promoting a team-based, non-adversarial approach among clients, advisers and the supply chain
- adopting a partnering approach wherever possible
- identifying risk and how best to manage it
- collecting and interpreting data on the performance in use of their construction solutions for the purposes of feedback.

Although, during the 1980s and 1990s, much of the innovation was essentially led by private sector clients, more recently central government and local authorities have been playing significant and increasing roles in championing and supporting innovation in construction. Since the late 1990s, HM Treasury and OGC Guidance on Construction Procurement has been encouraging its governmental departments to lead change in construction through a series of guides. This series of documents provides best practice advice at a strategic level and covers the clients' role in the construction procurement process, with particular emphasis on:

- roles and responsibilities
- training and skill development
- achieving value for money and
- project management.

They also promote and provide guidance on partnering and incentivisation. They are deliberately aimed at encouraging a change in the government client culture. The Fairclough report (2002) reaffirmed the important roles for government in improving the effectiveness of construction, as a major regulator, sponsor and client. As a regulator, government has a responsibility to establish a framework that anticipates emerging needs but protects minimum building standards. As a sponsor for the industry, the report argues that government should facilitate change but not impose or assume control. As a client, Fairclough sees the government as having a vital role in stimulating innovation by demanding better value and fitness for purpose from public buildings. This should focus particularly on the interests of the end users of buildings.

Other champions of innovation include other participants in construction's operating system such as main contractors, prime contractors, consultants, subcontractors and suppliers. Main contractors are well placed to provide a holistic view of the whole value chain because of their position between upstream and downstream processes. They can help link the customer to the supplier and negotiate, define and agree the customer's requirements. Using this approach means the output of the relationship can be determined and measured. As both the customer and supplier have negotiated and defined the requirements, they also know whether the output meets the initial requirements. Another benefit of this approach is that main contractors can facilitate and encourage frequent and direct communication upstream and downstream in the process. This can provide a basis for the development of trust, continuous improvement in reducing waste and adding greater value to the external customer.

Case study 8.5: Wates – main contractor championing innovation

As part of their Action Programme, Wates are investigating ways of improving relationships with their specialist contractors. They have undertaken a major survey to evaluate the effectiveness of their relationships with specialist contractors, and to identify ways in which they can enable their specialists to add more value for the external customer.

In 1995, Wates decided to introduce radical changes in their culture in order to improve profitability, provide non-linear growth in operating margins, and add more value for the external customer and end-user.

The intial focus was therefore on the need to change their own culture. Given the size of the organisation, its strong emphasis on 'command and control' and its anticipated resistance to change, it was decided to bring in an external consultant to help them create a transformation in their management techniques. This new culture is based on improving relationships and developing trust through removal of barriers, more effective communication, addressing the causes of conflict, and stimulating on-going learning. Disseminating the new culture throughout the whole organisation has proved to be a long and difficult task.

There was also some degree of realisation that this new culture needed to transcend the boundaries of the organisation and engage with other key players in the construction process. This is why, in 1997, they commissioned a survey to investigate the nature of their relationships with their specialist contractors. As a consequence of the results of this survey, Wates has been actively investigating the implementation of SCM as a means of increasing customer focus both up and down the construction process in order to differentiate themselves from their main competitors. For example, they have substantially reduced their customer base to increase upstream customer focus. They have also been encouraging customers to shift from price competition and lump sum contracts to two-stage, negotiated contracts. In addition, Wates has attempted to use these improved downstream relationships to play a leading and pivotal role in aligning and integrating both parts of the construction process.

As mentioned by a senior manager of Wates, this approach has helped to improve their business results. However, there are still substantial barriers to overcome including a lack of champions throughout the whole organisations, the continuing blame culture and the resistance of individuals to change – all of which are detrimental to forming better relationships with other organisations throughout the supply chain.

As a substantial proportion of Wates's future workload will be awarded on the basis of lowest price, there is an awareness of the need for the company and its supply chain to be both lean and agile in order to be well prepared to deal with the two main types of construction market (two-stage negotiated and price-competitive). Being lean is essentially about close collaboration, reducing waste and adding value through aligning and integrating processes. Agility is necessary to respond effectively to price-competitive projects and the needs of the irregular client.

Top-level leadership within project teams and their supply chains needs to be complemented by leadership at all levels and stages if innovation is to be diffused and successfully implemented and sustained. Diffusion is the process by which leaders ensure that the innovation is spread and assimilated by individuals and groups within their organisations, projects and supply chains. However, it is important to recognise that this diffusion process can take considerable time given the resistance to and fear of change, adversarial relationships, lack of learning and the blame culture that exist in much of construction.

Learning to support innovation

Learning refers to the various processes by which skill and knowledge are acquired by individuals and organisations and is increasingly being recognised as a key to innovation and change. One of the main reasons for this increasing interest in learning is the growing pressure on organisations to respond to the challenges of the rapid pace of change and global competition. The pressures and opportunities for change demand innovation, the success of which depends on learning (see Fig. 8.2).

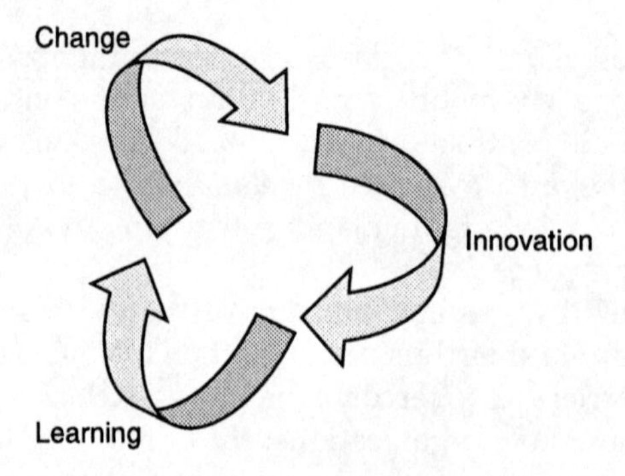

Fig. 8.2. Learning as a key to innovation and change

Three main types of learning have been identified which reflect the degree of certainty in which organisations are operating (see Fig. 8.3).

Single-loop learning corresponds to an era of relative predictability and certainty. It is the first and least demanding learning process which consists simply of adapting behaviour towards the attainment of existing goals. The organisation reacts by correcting errors within the norms and values which form the organisation's rationale. The existing norms and values are not questioned and they remain directed towards the existing purpose of the organisation.

With greater uncertainty and less predictability within the business environment, double-loop or deeper learning associated with behavioural adaptations becomes necessary. Significant changes occur in the relationship between the organisation and its environment necessitating more than a simple process of adaptation. Organisational norms and values

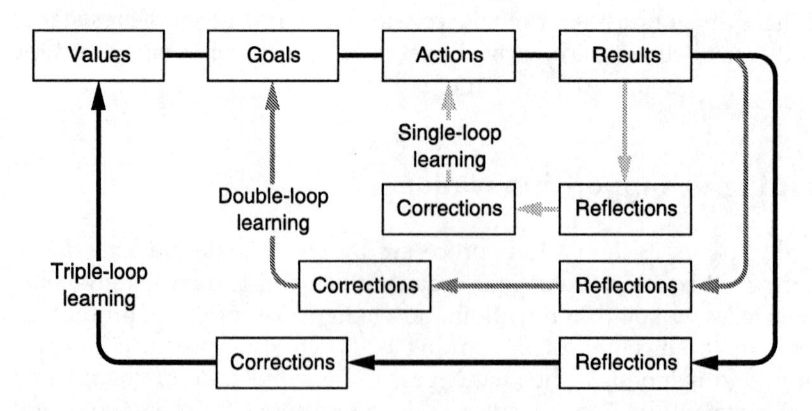

Fig. 8.3. Types of learning (derived from Argyris, 1992)

must be questioned and new priorities set which may well lead to modifying the organisation's goals and restructuring its value systems. Vital elements of this process are the open provision of information and the unlearning of learning cycles, neither of which are easy to achieve.

Triple-loop learning or learning how to learn is increasingly used to meet the challenges of the new paradigm characterised by rapid, significant and complex change. It corresponds to the highest level of learning. The aim is not to learn particular items of information or knowledge but to learn about the process of learning itself. Learning at this level involves reflection, analysis and the creation of a framework of meaning. It implies the unfolding and understanding of the learning process itself. When an organisation learns to learn, its relationships and environment are more clearly understood and the true role and meaning of the organisation are revealed. This can lead to fundamental transformation of the organisation.

There are many organisations which experience great difficulty in matching their type of learning to the level of complexity they face and move beyond single-loop learning. The relationship between the types of learning and the degree of complexity is shown in Fig. 8.4.

For example, when a radical transition is necessary, they prefer to let a third party determine how it must be made instead of mobilising the knowledge and ability often present in abundance within their organisation. Another common mistake is to start up new learning processes without exploiting the previous learning by using it to effect visible and concrete behaviour change. Organisations which begin to recognise these issues for themselves and which want to examine them and draw lessons from them, are organisations which are learning how to learn. This can lead to the recognition that the ability to learn faster than com-

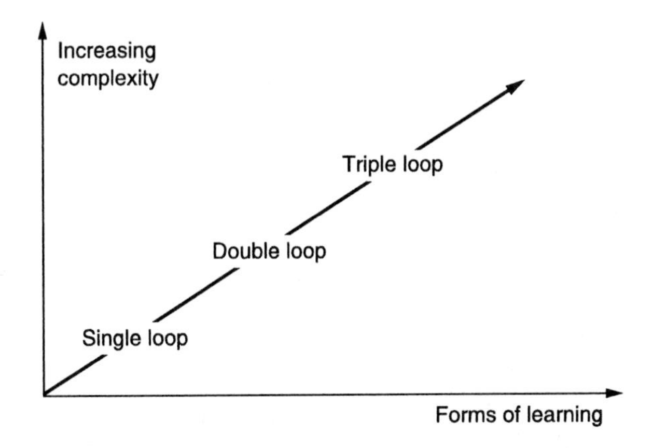

Fig. 8.4. Type of learning versus degree of complexity (Jones and Saad, 1998b)

petitors may be the only sustainable competitive advantage. Learning is not seen as an indulgence but a competitive necessity as people learn through working together and work together through learning (Garratt, 1990). It is not an isolated or occasional exercise but a continuous activity which facilitates a continuous adaptation to changing circumstances through challenging traditional values, norms and learning approaches. It can tap the individual's commitment and capacity to learn at all levels in the organisation. It affects and engages everyone's learning to create a whole which is greater than the sum of its parts.

There is therefore a dynamic and complex relationship between the changing environment and the learning process. To increase effectiveness and competitiveness the pace of learning must be equal to, or even greater than, the rate of change being faced. In this post-Fordist era, characterised by globalisation and fierce competition, where the only certainty is uncertainty, the one sure source of competitive advantage is knowledge (Nonaka, 1996). The most successful companies are those which constantly create new knowledge, disseminate it widely throughout the organisation and quickly embody it in new technologies, products, pratices and ways of working. These activities define the 'knowledge-creating organisation' whose primary aim is continuous improvement.

The interest in the concept of organisational learning grew during the 1990s with the emergence of post-Fordism and its emphasis on the continuous acquisition and renewal of knowledge and its dissemination within both organisations and networks such as supply chains. Organisational learning describes attempts by the whole organisation to promote learning in a conscious, systematic and synergistic way. Its major objective is to enable the whole organisation to enhance its innovative capacity and hence its performance and competitiveness through the acquisition of more relevant and up-to-date knowledge and understanding. Organisational learning takes place in and through interaction with and between a number of people. Clearly, an organisation can only learn because its individual members learn. Without individual learning there can be no question of organisational learning. However, individual learning is a necessary but not a sufficient condition for organisational learning. An organisation learns if someone not only does a better job but, as a result of this, other members of the organisation operate differently. Thus, individual learning needs to be captured and transferred into collective or organisational learning. Organisational learning involves basically the development and application of both. In line with Kolb's learning cycle (1975, 1976) and as illustrated in Fig. 8.5, organisational learning is a dynamic process built around the idea of continuously questioning, revising, interpreting, problem-solving and creating and integrating new knowledge in an organisational context (Dixon, 1994; Revans, 1983).

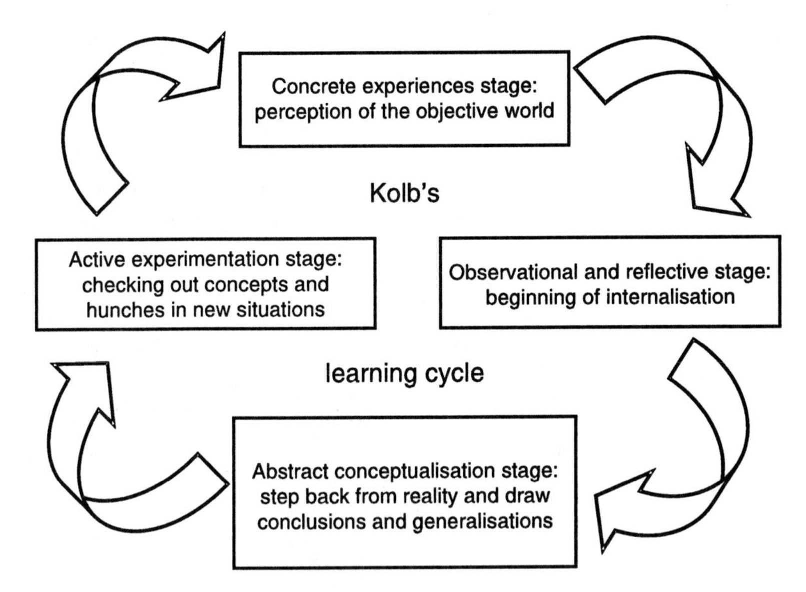

Fig. 8.5. Kolb's learning cycle

The concept of 'organisational learning' has also led to the concept of 'management of knowledge' or 'knowledge management' which is often used to describe the management routines that can help people in organisations share what they know. It is a means to acquire, disseminate and exploit knowledge within organisations in order to maintain and improve competitive advantage through innovation. The longer-term challenge is to facilitate the transformation of construction organisations into learning organizations as shown in the Guinness Trust case study. The second challenge for construction organisations is to increase the scope of learning through sharing information and knowledge between organisations and hence developing inter-organisational learning.

Case study 8.6: Guinness Trust – organisational learning

Housing Association Guinness Trust runs in-house conferences four times a year. These range between one and four days' duration. These focus on *Rethinking Construction* and other important issues. The events are used to keep development and maintenance staff from around the country informed about progress on new initiatives. They also provide an opportunity for consultation and the sharing of learning. Both in-house staff and external experts facilitate these events.

Achieving organisational and inter-organisational learning is difficult. It is often impeded by the following factors:

- learning is often seen as undermining authority and power within and between organisations
- deeply embedded adversarial relationships and lack of trust
- vulnerability associated with the process of unlearning needed for cultural change
- lack of flexibility and unwillingness to risk something new and learn from mistakes
- lack of leadership and commitment to learning and experimentation
- uncertainty and ambiguity that learning can create
- lack of skills including listening, coaching and facilitating skills
- lack of time and resources
- arm's-length dealing
- short-term and transaction-based relationships
- tight boundaries
- tradition of confrontation and blame culture
- lack of customer involvement.

Inter-organisational learning is strongly dependent on effective inter-organisational relationships which require the following changes:

- close and cooperative links
- long-term relationships and development
- blurred boundaries with extensive and multi-level crossings
- partnership in mutual development
- seeing the customer as an integrated part of the business.

However, there are significant problems when sharing learning across the boundaries between organisations including differences in:

- aims and values
- language and means of communication
- practices and procedures
- culture
- technology and communication systems.

In addition, issues of power, organisational structures and levels of bureaucracy are also major barriers to sharing learning between organisations. This challenging type of learning requires new ways of thinking and changes in the organisation in terms of structure, leadership, communication and culture. It is clear that boundaries, bureaucratic procedures and adversarial relationships work naturally against the organisational and inter-organisational learning as they impede participation, involvement, cross-learning and innovation. Therefore, organisational structures need, for example, to change from being vertical, hierarchically structured, functionally-oriented and fragmented to becoming flatter, more horizontal, cross-functional and integrated.

Layers of management provide another obstacle to motivation, participation, learning and innovation. De-layering is a popular tool of organisational transformation today which is strongly linked with people empowerment in order to release their energies and place decisions closer to the point of impact. Reduced layers means reduced filtering and greater availability of information, and greater sharing. It frees people for more interesting, effective and creative tasks within and between organisations.

However, it is important to recognise that construction projects involve a large number of independent but inter-related organisations. This implies that improvement across the whole project process requires learning across organisational boundaries and a sharing of learning. This opens up different opportunities for capturing and exploiting learning involving all participants in the project: individuals, organisations, the project teams, supply chain alliances, professions and trade associations. The possibilities for the spread of cross-network learning are limited because of the very small number of construction clients currently adopting innovations such as supply chain management. In addition, this learning should focus on technical aspects as well as the inter-personal skills which are needed to develop and sustain the new culture. However, learning in construction is less formalised and prioritised than in other sectors.

In addition, transfer of knowledge from one party to another is never effortless and instantaneous, not even if both operate in the same environment. Knowledge has environment-specific elements. It is always a product of the economic, climatic, social and cultural context within which it was designed, as discussed in Chapter 4. Efficient use in an environment that is different from the one in which the knowledge was developed will generally call for modifications and adaptations. Overcoming tacitness and making environmental adaptations will thus generally entail the need for a new culture, appropriate organisational structure, resources, commitment and new or enhanced skills to assimilate effectively and adapt this knowledge to local conditions, improve upon it and, ultimately, create new knowledge.

External support

The capacity of construction organisations to innovate depends not only on their internal capabilities and those of their partners within project and supply chains but also on the support they receive from the external environment. As shown in Fig. 2.1, support institutions are part of the external environment. In an innovative milieu there is interaction between organisations and the support institutions in order to coordinate the direction of information gathering, learning and R&D efforts. Links with research and teaching institutions can lead to greater understanding of the main prerequisites for a successful implementation of innovation (Saad, 2000; Tidd *et al.*, 2001).

Case study 8.7: The Movement for Innovation

In October 1997, the Construction Task Force, chaired by Sir John Egan, was commissioned to advise the Deputy Prime Minister from the clients' perspective on the opportunities to improve the efficiency and quality of delivery of UK construction, to reinforce the impetus for change and to make the industry more responsive to customer needs.

During July 1998, the Construction Task Force published their report *Rethinking Construction*. It contained the clear message that the industry would not significantly improve unless it embarked on a course of radical change. This would involve a totally new approach to the delivery of the construction product.

The report identified committed leadership, a focus on the customer, integration of the process and the team around the product, a quality-driven agenda and commitment to people as the key drivers required to implement change and develop substantial improvement to the project process.

The project process can be radically improved by fostering innovative methodology. The innovations would be categorised under the headings of product development, project implementation, partnering the supply chain and production of components.

The Construction Task Force also highlighted the need for industry to set clear, measurable objectives and to create a performance measurement system to aid benchmarking and provide tools for sustained improvement.

With these changes the Construction Task Force believes that the industry can achieve annual improvements of:

- 10% reduction in capital cost and construction time
- 20% reduction in defects and accidents
- 10% increase in productivity and profitability
- 20% increase in predictability of project performance.

The report proposed the creation of a 'movement for change' which would be a dynamic, inspirational, non-institutionalised body of people who truly believe in the need for radical improvement within the construction industry.

The Movement for Innovation (M4I) was subsequently launched on 3 November 1998 to facilitate this cultural change. It aims to lead radical improvement in construction in value for money, profitability, reliability and respect for people, through demonstration and dissemination of best practice and innovation.

Its strategies to deliver these goals are as follows.

- To bring together clients and all involved in the construction supply chain, in innovation, best practice, or research, who are committed to change and innovation in construction.
- To provide leadership, share experience and work together to create an open, cooperative, no-blame, non-adversarial team approach to innovation.
- To drive forward by example and persuasion the changes needed to create an industry in which the norm will be committed leadership, a focus on the customer, a process and team integrated around the product, a quality-driven agenda and a commitment to and respect for people.
- To facilitate delivery of the enhanced performance targets set out in *Rethinking Construction* through sustained improvements and innovation in product design and development, in project implementation, in partnering the supply chain and in production of components.
- To test, measure, quantify and disseminate experience and achievements from demonstration projects in the form of case histories, toolkits, guidance notes, themed events and conferences.

The M4I's principal achievements include the following.

- The development of the industry's first ever set of standardised Key Performance Indicators. For the first time the industry and its clients are able to measure key aspects of performance, to set and monitor progress against improvement targets and to benchmark against the best in class. These measures are being used to determine performance of the industry at large and specifically to measure the business and organisational benefits from the M4I Demonstration Projects. As a result of the Board's efforts, measurement using standard tools is increasingly part of mainstream business practices within the industry including the integration of the project team and long-term partnerships.
- The Board is actively supporting the work of the Confederation of Construction Clients and their proposals for a Client's Charter. Client 'pull' in support of innovation, best practice and performance measurement has a vital role to

play in supporting continuous improvement in performance of the industry and its clients.

- In support of the vital Recruitment, Retention and Respect for People agenda, an M4I Working Group has produced a report to ministers on the way forward which incorporates six toolkits to monitor performance and improve key aspects of performance. The construction industry is in competition for the best people and must commit itself to achieving significantly sustained improvement in performance if it is to recruit, retain and get the best from its people.
- An M4I Working Group on sustainability has developed a set of Performance Indicators and is working on a set of three toolkits which will be available for use early next year to help improve the sustainability of construction. The Board is also working with the Commission for Architecture and the Built Environment and other organisations to promote the contribution that good design makes to an enhanced urban environment.
- The Board has developed a broad based programme of Demonstration Projects. At the time of writing 170 Demonstration Projects are in management. To further extend the coverage of the programme they are working to recruit additional Demonstration Projects from manufacturers, specialist contractors, clients, designers including, in particular, nominations from Small and Medium Enterprises (SMEs).
- The Board has now published 41 Case Histories detailing the proven innovations from Demonstration Projects. Additional Case Histories will be published at regular intervals. All the Case Histories are being shared with industry outside the Demonstration Project Programme through the Construction Best Practice Programme.
- They have established a network of nine Regional Clusters through which the Demonstration Projects meet regularly to share information on their project innovations and best practices. These Regional Clusters are increasingly expanded in their network of contacts with other regionally based organisations seeking to promote *Rethinking Construction* and other business improvement.
- Through joint membership at Board level and joint work programmes the Board is actively collaborating with the Housing Forum, the Local Government Task Force, the Government Construction Client's Panel, the Construction Best Practice Programme and with other organisations working to support the implementation of *Rethinking Construction*.

- In response to requests from industry they have recently launched a membership scheme to enable organisations not currently involved in Demonstration Projects to play a direct part in M4I activities. The membership scheme is likely to be particularly important in facilitating increased involvement by SMEs and regionally-based organisations.
- The progress that the UK has already achieved in engaging widespread support to radically improve the performance and value for money provided by the construction industry is attracting significant international interest from government and industry organisations.

As discussed above and in Chapter 3, there is increasing external support, at both local and national level. However, it is not reaching most of construction and is not totally reflecting the learning needs associated with the level of complexity characterising post-Fordism. In addition, this support is based on a universal best practice which fails to take into consideration the specificities of construction sectors and regions. Morgan (1997) and Maillat and Kébir (1998) propose the notion of the learning region and emphasise that innovation is an interactive process between an organisation and its local environment which can create an innovative milieu which in turn can foster, stimulate and shape innovation. This is being addressed by the regionalisation of the *Rethinking Construction* movement.

Case study 8.8: Rethinking Construction

The Rethinking Construction Board comprises a ministerial steering group chaired by the Minister for Construction. It oversees initiatives launched by the 'four strands' responsible for implementing the principles of *Rethinking Construction*.

The Movement for Innovation leads on non-housing construction in the private sector. The Housing Forum leads on housebuilding, refurbishment, and repairs and maintenance in the public and private sector. The Local Government Task Force leads on best practice for local government clients. The Central Government Task Force leads on best practice for central government clients.

Partner organisations include the Construction Best Practice Programme and the Government Construction Clients' Panel. The former raises awareness of the benefits of best practice, and provides the construction industry and its clients with the skills and knowledge to implement change and the latter works to ensure that all government clients have the skills to become and remain best practice construction clients.

Case study 8.9: Rethinking Construction at a local level – the case of the South-West

To spread awareness of the *Rethinking Construction* messages to the widest possible audience and to engage more effectively with regional clients, contractors and suppliers' action is underway to develop a network of Rethinking Construction Centres across the UK.

In collaboration with organisations already working locally to promote the business and organisational benefits of *Rethinking Construction*, the intention is to establish at least one Rethinking Construction Centre in each of the English Regions (within the boundaries of the Regional Development Agencies) and one each in Scotland, Wales and Northern Ireland.

Centres in Northern Ireland, Wales and the North-West have already been established. The target is to complete a national network of twelve centres by the summer of 2003.

The precise programme of work for each of the centres is being determined locally in accordance with local needs and priorities. However, the principal tasks include:

- providing a mechanism to enable organisations and individuals actively supporting change to communicate with each other
- helping to communicate the business lessons from *Rethinking Construction* to regional and local organisations
- supporting the continuity and local ownership of the *Rethinking Construction* agenda over the long term.

Rethinking Construction South-West

A meeting of those interested in supporting the setting up of a Rethinking Construction Centre in the South-West took place in July 2002. Almost 50 representatives of organisations including South West Regional Development Agency (SWRDA), universities, clients, Construction Best Practice Clubs (CBPC), CITB, CIC, regional professional organisations, consultants and contractors attended.

A 'shadow' Board met in September and in early November 2002. A number of task groups have been established to engage with skill shortages and related issues, procurement methods, sustainable construction and promoting best practice. The proposal for a Regional Rethinking Construction Centre is strongly supported by SWRDA in part because of the fit with the Office of Deputy Prime Minister (ODPM)-led initiative 'Creating Excellence in Urban Regeneration'.

A new culture conducive to learning and innovation

The effective development of new values to encourage participation, commitment, involvement, motivation and learning is associated with the establishment of a supportive organisational culture. Culture, as defined in Chapter 3, means the personality and behaviour of the organisation. Culture refers to a system of shared meaning held by members that distinguishes a group of people or an organisation from other groups or organisations. The system of shared meaning is a set of key characteristics indicating an organisation's current customs, traditions, and its general way of doing things. It therefore informs workers how things are done and what is important.

Cultural change is a process whereby new meaning emerges, leading to the reinterpretation of past experience in order to promote new lines of thinking and actions likely to encourage motivation, creativity and innovation. It is a careful re-engineering of the 'fabric' of the organisations and reshaping of the way things are done aimed at repeating and improving what works and giving up what does not. It is therefore a powerful determinant of people's beliefs, attitudes and behaviour. It can help to bind people together through a sense of belonging and a sense of common purpose and can also generate an environment conducive to competitiveness through joint learning and innovation.

When new challenges occur, or when the environment is undergoing rapid change, the organisation's established culture may no longer be appropriate and can be an obstacle to the organisation's effectiveness. There is therefore a vital need to develop a new set of norms and values associated with effective participation, sharing of information, joint problem solving, open communication, joint learning and cooperation. Once these new lines of beliefs, norms, conducts and actions are articulated and described, they need to be learnt by the organisation's members (individual learning) in particular and the organisation in general (organisational learning).

A culture is learned and therefore its values have to be taught. This applies to the cultural values of an organisation as it does in the context of school, family or society. Values only have power to control and direct minds and behaviours when they are internalised. They will only be internalised when they are accepted and when they actually mean something to the individuals concerned. Values are the underlying beliefs and attitudes that help determine conduct. They shape motivation and motivation delivers outcomes. The Pearce Retail case study shows an early attempt to use the People Performance Indicators (PPI) to promote dialogue on the 'soft' people issues that promote more proactive attitudes and behaviour.

Case study 8.10: Pearce (Retail Services) (1) – use of People Performance Indicators (PPIs) to promote attitudes and behaviour conducive to innovation

Pearce (Retail Services) are demonstrating how PPIs and their toolkit measurement system can provide a catalyst for improvement. Charlotte Curtis, Performance Measurement Coordinator at Pearce explains why these tools were needed. 'As part of the formal measurement system that we use, we had been looking for measures that would help us improve the business by focusing on people. Getting effective measurements of relevant 'people' issues was the key. The checklist and scorecard format offered the answer.'

The PPIs and four toolkits were introduced on two quite different projects – a large extension in a busy city centre and the refurbishment of a building on a retail park. The checklists were used to interview personnel on both sites from Pearce, specialist and trade subcontractors and clients. Scorecards were then completed on the basis of the answers given which were used to identify key competitive strengths as well as areas for development. A working party drawn from across the business identified where improvements could be made and implemented the necessary changes to processes, roles and responsibilities.

The approach has helped Pearce to address areas which are often hard to quantify. It has encouraged people to think about stopping bad practices in relation to health and safety. The toolkits were also used to better understand the causes of stress on projects and how they might be tackled. Issues that need to be covered in site inductions such as cultural diversity, gender awareness and customer care are highlighted by the toolkit.

Curtis explains that further developments include 'integrating some of the toolkits into our measurement system to be rolled out across all projects, using the Worker Satisfaction toolkit with our direct and indirect employees. We'll need to repeat the measurements periodically to check that we're making progress. The Respect for People KPIs will remain an external reference point.'

Movement for Innovation, Case History, www.m4i.org.uk

When leaders and managers only pay lip-service to the values they preach and are found out, then such hypocrisy creates cynicism, scepticism and ridicule. It undermines motivation and creates barriers to learning and innovation. No human beings can exist without values. If values are not clear, a counter-culture can emerge that will have its own system of values and principles and codes of behaviour.

It is, however, important to point out that the attribution of a culture to an organisation is a relatively recent phenomenon and insufficient empirical research has been carried out so far to examine the relationships between organisational effectiveness and organisational culture. Organisational culture is vast and complex and includes a whole range of intangible factors, including style of management, preferred goals and preferred means of achieving them, types of people who belong to the organisation, particular traditions or myths from the organisation's past and the legacies of particular, strong personalities. The second case study based on Pearce Retail is also aimed at creating a culture more conducive to innovation.

Case study 8.11: Pearce (Retail Services) (2) – using Team Climate Inventory (TCI) to create a culture conducive to innovation

Pearce Retail use the Team Climate Inventory (TCI) used to measure team performance on a project for Safeway in Chelsea. The method provides a means to evaluate the shared perception of how people feel about decisions, communications, practices, etc. TCI shows its users where weaknesses exist through the use of a questionnaire that may be completed by team members as frequently as they deem necessary. It asks about the atmosphere in the team including:

- how people tend to work together
- how frequently they interact, their objectives, and
- how much practical support is given towards the implementation of new and improved ways of doing things?

Pearce Retail borrowed the idea from the oil industry and NHS management teams. They argue that the industry needs more hard and soft measurements if it is to raise its game, and that means gathering accurate data and acting on it. The TCI provides a 'soft' measure in relation to behavioural change and attitude development. It helps to communicate the team vision – the team has a clear sense of purpose and understands its collective strengths and areas for improvement. It provides structured feedback for team members on their performance and on the project atmosphere based on their aggregated self-assessments. It encourages innovation – for example, the Safeway Chelsea team needed a radical solution to overcome a logistics headache at this congested inner city site. Assessing the team climate inventory particularly with the site foremen in the pre-start workshop really got the whole team working to solve this problem.

Movement for Innovation, Case History, www.m4i.org.uk

An evolving model of the diffusion of innovation in construction

Given the fundamental changes associated with complex innovation, coupled with the nature of the construction process and its project-based approach, it is appropriate to begin diffusing innovation at the project level. The choice of the project as the outset of the diffusion process is essentially motivated by two reasons. First, the project is seen as the most appropriate and readily available context within which construction organisations can be influenced and supported in their attempts to embrace innovation. Second, experiences from other sectors suggest that commencing the implementation of innovation at the shop floor, or operational level, is likely to be more successful. In the context of construction, the project is seen as corresponding to the shop floor level (Jones and Saad, 1998a). The first suggested step in the implementation process starts with the project. This is illustrated in Fig. 8.6.

This is leading to construction organisations transferring best practice from the project to their own company. They can adopt innovations learned within the project to improve their company's performance. This diffusion process from projects to construction-based organisations is illustrated in Fig. 8.7.

Having capitalised on the changes within projects and companies, the next logical action is to extend learning and best practice beyond the

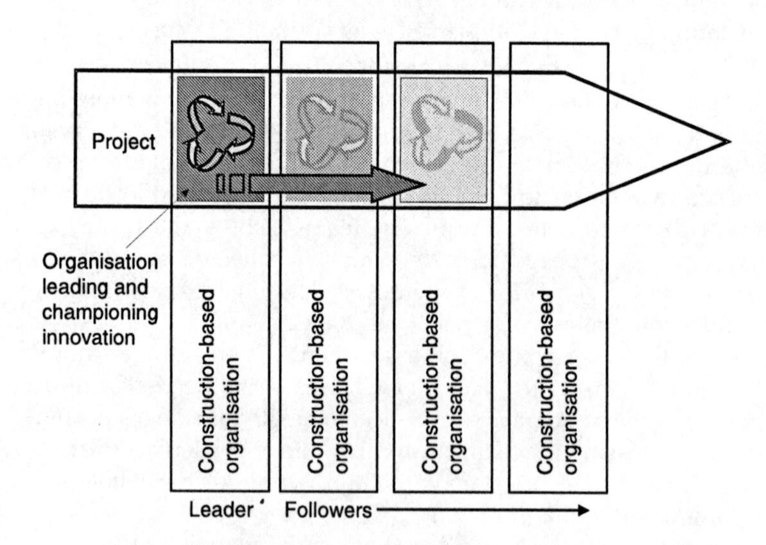

Fig. 8.6. Diffusion innovation at the project level

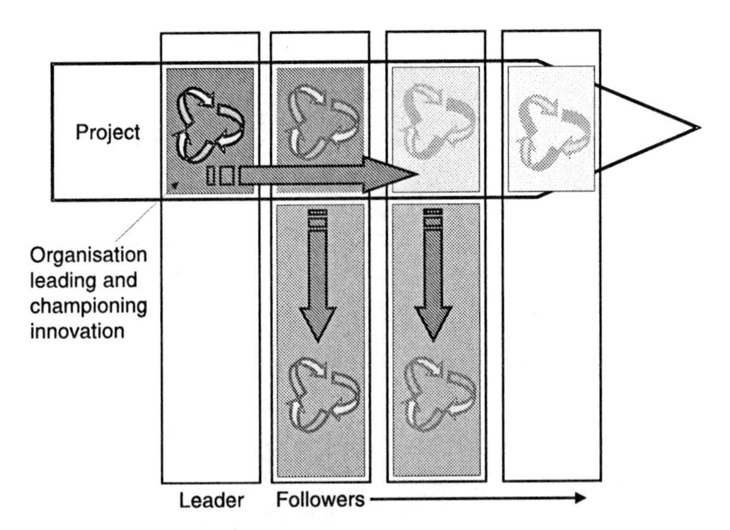

Fig. 8.7. Diffusion of innovation at the organisational level

boundaries of individual projects. This is taking place as construction organisations become involved in other projects and disseminate their learning and best practice to other project teams.

As has been seen in previous chapters, more significant and sustained innovation in construction is also being achieved through the development of closer, longer-term strategic relationships based on partnering and supply chain management, as shown in Fig. 8.8. In some sectors of construction this has resulted in dramatic improvements in performance and competitiveness. Such collaboration has also resulted in greater equality and trust between partners as they work together in the longer term to fully satisfy customer needs and improve mutual competitiveness.

Having begun to implement these changes within clusters of companies and projects, the remaining challenge is to diffuse the best practice associated with this approach throughout the whole construction sector as illustrated in Fig. 8.9.

In this way, the whole culture of the industry will to continue to change and this will also help meet the requirements of the irregular customer whose needs may not have been addressed through the improvements introduced in strategic partnering. As explained in Chapter 3, national strategies for the implementation of innovations to develop a world-class construction industry were set up following the publication of the Latham report in 1994. The Construction Industry Board was set up to drive forward the changes recommended in the report. Again, following the publication of the Egan report in 1998, there has been significant and growing support for construction organisations in responding to the new paradigm.

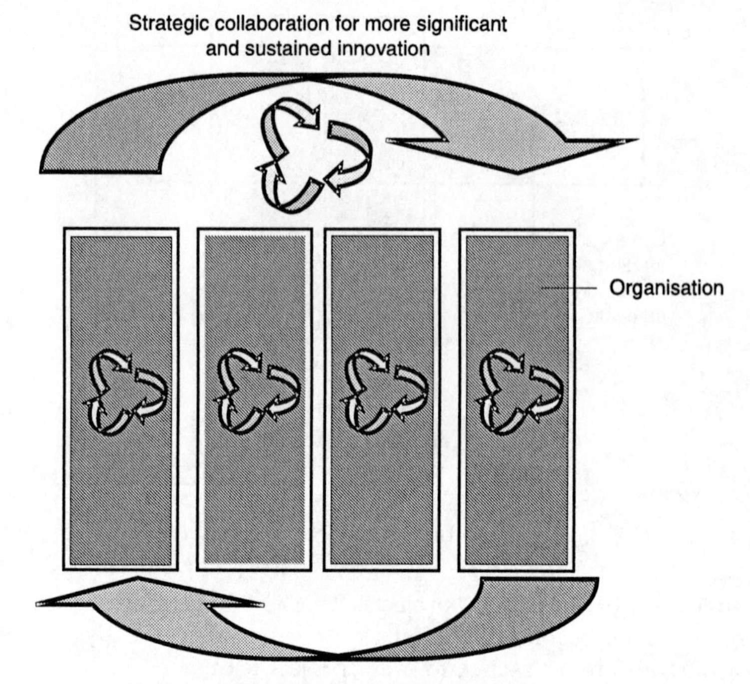

Fig. 8.8. Diffusion of innovation at the strategic level

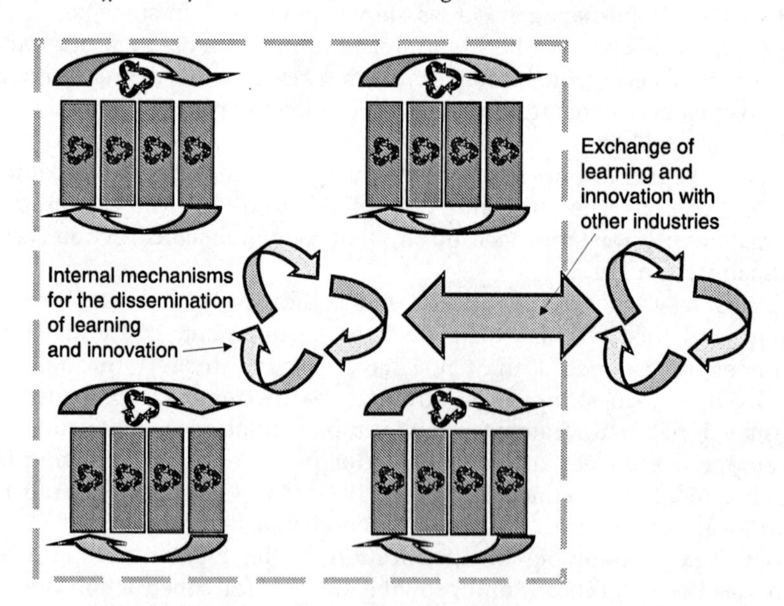

Fig. 8.9. Diffusion of innovation throughout the industry

Case study 8.12: The Construction Research and Innovation Strategy Panel (CRISP) champions research and innovation

CRISP is operating closely with the Strategic Forum for Construction and others to promote the benefits of research and innovation to the whole construction community, and to identify key Research and Innovation (R&I) priorities at a strategic level.

CRISP considers that the current climate of change provides a window of opportunity for construction companies to improve their competitiveness through R&I.

This book has attempted to identify and examine the main determinants of innovation in construction, which emphasise a more holistic and collaborative approach. However, in the case of the majority of construction organisations, a successful implementation of innovation is strongly dependent on their internal receptiveness to and preparedness for this long and difficult journey as demonstrated in the Waterloo Air case study.

Case study 8.13: Waterloo Air Management (WAM) – developing an internal culture as a prerequisite to more collaborative ways of doing business in supply chains

WAM is one of Europe's longest established manufacturers of bespoke ventilation products for the construction industry including a comprehensive range of heating, ventilation and air conditioning (HVAC) equipment. This comprises air terminal devices; fire, smoke and control dampers; and air handling and filtration systems. WAM also supplies ranges of products from other manufacturers to meet its customers' requirements. It has extensive manufacturing and testing laboratory facilities at Maidstone in Kent and Holten in eastern Holland with a combined workforce of around 250 employees and a turnover of over £20 million.

Up to 1996, WAM essentially had two businesses – one providing domestic ventilation products and the other commercial ventilation products. Although the commercial business was not profitable, the businesses together generated a small profit. In 1996 it sold the profitable domestic business and began focusing on running the commercial business and making it profitable. In determining a new profitable strategy, WAM began to investigate what was going on in the outside world in construction and the wider environment. This corresponded with the national debate in relation to the performance of the construction industry, which had been promoted by the publication of the Latham report two years earlier in 1994.

In 1996, WAM's Deputy Chairman was appointed as Deputy Chairman of the Federation of Environmental Trade Associations. The Construction Industry Board (CIB) was just being set up and he quickly found himself as a member of its Board even though at the time WAM's turnover was only £9 million. In fact, it was the only manufacturer who volunteered to join the CIB. From these early connections with the wider industry, WAM's Chairman was then invited to join the Construction Best Practice Group. Then, in 1998, the Egan report was published along with its implementation initiatives, which included the Construction Best Practice Programme. As WAM was still the only manufacturer contributing to the debate regarding the future of construction, it was invited onto that too. This means that WAM has gained a reputation as being at the leading edge of change in construction – although paradoxically its people do not see themselves as anywhere near the leading edge when viewed from other sectors of the economy.

The important outcome of this greater engagement with construction was that WAM began to listen to and understand its customers. Also, for the first time ever the team members were allowed to discuss and challenge the long established ways of doing business in construction and contribute equally to the debate concerning the future direction of construction in a way that had never been possible before. This was a radical change because traditionally construction was seen as 'being incredibly deaf in terms of what goes on outside individual businesses'.

This greater engagement with construction persuaded the management team that they would take their business and make something different of it in the new post-Latham construction environment, which is what they have been doing during the five years since January 1997, when they launched their new business plan. This involved a five-year programme of internal transformation, much of the credit for which is attributable to the chairman Ron Edmondson who had the vision in the mid-1990s to recognise the potential of more collaborative working in supply chains.

When it started out on this programme of change it wasn't partnering that was driving WAM but, essentially, customer service. Increasingly, WAM realised that the way it served its customers and how its customers served their customers in construction was 'awful'. In many aspects of life, more and more people know what constitutes good service in a shop or a restaurant, and increasingly WAM's construction customers were expecting the same level of service. Price used to be the main deciding factor in the selection of manufacturers but increasingly customers have begun choosing to work with WAM because they are seeking value for money, which might not be the lowest price. It is a change that has affected everything and everyone in the company.

In this more customer-focused approach, the attitude of people in WAM and its culture were seen as being as important, if not more important, than its products. The rationale underpinning this approach was that this is where real value and business benefits are to be found. In the past, WAM had placed the onus for flexibility onto its customers with customers having to be flexible to fit in with the way manufacturers operated. WAM considered this was not the right way to treat customers, so it has worked to provide a flexible service for its customers rather than the other way round. Increasingly, WAM is providing the flexibility in its own organisation and processes rather than passing that responsibility onto its customers. The company has become much more responsive to demand or 'pull' from its customers and this has been the real key to its success.

WAM quickly recognized that its new customer-focused strategy needed to be supported and sustained through an intensive internal programme of education and training. A Charter was developed to drive forward an internal change programme by clearly setting out the key standards of performance for WAM's business and the world-class standards that customers can expect. These were produced to educate WAM's own staff on how to deal with their customers, including how they should answer the phone, deal with problems, how they turn round quotations, but, most importantly, how they deliver. The industry delivery standard for grilles and diffusers is 8 to 10 weeks, which WAM has reduced to 5 days for smaller orders and up to 10 days for bigger orders. WAM sees these levels of performance as being unique in the industry and setting the standard for the entire market. These standards are detailed below.

- *Quality*: registered to BS EN ISO 9001:1994, WAM strives to achieve 100% quality with all its products. It recognises that quality requires constant vigilance and monitoring, not only of products but also of internal processes and systems.
- *Service*: in all dealings with customers WAM's employees aim to be prompt, courteous, and friendly and to fulfill their promises. They set delivery times based on defined standards and aim to deliver on the day and at the time they commit to.
- *Value*: WAM pledges to work with its customers to achieve the best combination of specification and cost to suit their requirements.
- *Problem recovery*: WAM recognises that even with the best run projects unexpected problems might arise and that it is

important they are sorted out swiftly. It has a no-fault or no-blame approach to problems and their solutions, which reflects its objectives of delivering a world-class service. WAM believes that any benefit gained in trying to apportion blame is far outweighed by the savings flowing from prompt solutions and the value of repeat business.

Equally as important as these key standards are its Client and Consultant Charters. The Clients' Charter covers levels of service of benefit to clients including issues such as value engineering and efficient project completion. Similarly, WAM's Consultants' Charter deals with levels of service and technical support, covering aspects such as mock-ups and verification testing, product sample supply and high levels of sales engineer response.

- *Value in design*: one of WAM's main strengths is its ability to meet the requirements of customers with unbiased and authoritative advice and the expertise to efficiently design, manufacture and test a wide range of products.
- *Development*: WAM has an R&D programme that includes significant investment in the design, testing and launch of carefully selected and specified products. Part of this programme aims to address to the complexity of the construction industry and to simplify the construction and installation process without compromising quality.
- *Project management*: WAM has a management system that allows it to design and deliver its products to construction projects – many of which can take many months to complete.
- *Testing and environmental facilities*: WAM sees product verification as remaining a fundamental element in providing high levels of customer satisfaction. It has a sophisticated testing laboratory that is used to undertake a range of aerodynamic, environmental and acoustic tests.
- *Investing in people*: WAM sees its people as being the most important part of the company and has been awarded the Investors in People award.

A number of improvements followed on from this approach such as 'no quibble', which is based on the approaches of most high street retailers such as Marks and Spencer. Customers like this type of service but it is also proving to be a money-saver for WAM. Surveys of rectification costs are showing that the cost of sorting out problems has dropped dramatically. This has been achieved because the 'no quibble' policy has forced WAM's people to get it right first time because it affects their bottom line. Also, they don't spend as much time on resolving problems – getting in and sorting things out. The emphasis is now on getting the next project from the customer rather than harrassing them for a problem on the current project.

What are the benefits flowing from this new customer-focused strategy? WAM is continuing to measure the benefits of its internal change programme. It has transformed losses of £2 million into profitability and more than doubled turnover. Radically improved delivery can be seen as the key to this success. In recognition of its new high standards of delivery, WAM is a member of the government-sponsored 'Inside UK Enterprise' initiative to demonstrate to construction firms how they can improve their performance and achieve high levels of customer satisfaction. Turnover has risen – in the case of grilles and diffusers, it rose from £2.5 million in 1996/97 to over £7 million in 2000/01.

WAM has also measured the benefits of its transformation using the Construction Industry Key Performance Indicators (KPIs). Together with Crown House Engineering, it has benchmarked progress in seven main areas by revisiting orders completed in 1996. The results are as follows.

- Purchaser satisfaction – WAM contends that measurements of credit notes as a percentage of sales (fallen to a third of 1996 figures) and overdue debt (now only 7%) indicate a dramatic improvement in customer satisfaction with products and service.
- Construction cost – many standard products are now cheaper than they were in 1996, a substantial reduction in real terms and the result of lean thinking over a period of five years.
- Construction time – average quoted lead times have fallen from about 40 days in 1996 to five days and WAM are now aiming for a day.
- Defects – WAM's no-quibble rectification policy has not only cut rectification costs by two-thirds, but also reduced the customer's costs of snagging.
- Predictability – fixed prices to key partners and improved on-time delivery (up from 40% to 95% over the five years) means that WAM's customers can price and programme with greater certainty.
- Productivity – the value of sales per employee has increased by 40%, even though the sales force has been halved.
- Profitability – WAM has broken its loss-making habit and profitability is on the ascent.

From the above, it can be seen that WAM has a growing reputation for customer service, innovation and product quality as it seeks to provide a world-class customer service based on its employees' attitudes and a belief in values such as flexibility, innovation, customer service and integrity. These, WAM maintains, are its guiding values that differentiates WAM from other organisations.

Key messages:

- do not underestimate the internal cultural change involved in adopting and sustaining more collaborative relationships and ways of doing business in the supply chain
- ensure that the commitment needed at all levels in the participating organisations is present
- do not underestimate the time it takes to effect real internal cultural change – having worked on this for five years WAM now thinks that the shift has been made
- recognise that more effective and customer-focused relationships in the supply chain affect everyone and everything in each company involved
- be aware that the changes will require a mix of hard (technical) and soft (cultural) changes
- use robust measures of performance in key areas of value to demonstrate progress
- recognise the challenges in making the changes stick, managing them and coping with growth at the same time.

The main factors necessary for a successful internal change programme are summarised in Table 8.2.

Table 8.2. Key factors for internal preparedness

- Clear and sustained goal to which people can commit
- Clarity of purpose for change
- Internal and external customer focus
- Performance measurement for continuous improvement
- Diffusion through champions at all levels of the organisation
- Open and effective communication
- Valuing people
- Motivation and empowerment
- Effective teams and leaders
- Appropriate organisational structure and culture conducive to learning and innovation
- Fast and early successes to increase momentum

References

ARGYRIS, C. *Organisational Learning*. Blackwell, London, 1992.

DIXON, N. *The Organisational Learning Cycle – How can we Learn Collectively?* McGraw-Hill, New York, 1994.

FAIRCLOUGH, SIR JOHN. *Rethinking Construction, Innovation and Research: A Review of Government R&D Policies and Practice*, 2002. DTI/DTLR.

GARRATT, B. Creating a Learning Organisation – A Guide to Leadership Learning and Development. Director Books, 1990.

JONES, M. and SAAD, M. *Unlocking Specialist Potential: a More Participative Role for Specialist Contractors*. Thomas Telford, London, 1988a.

JONES, M. and SAAD, M. Loop the loop to get ahead – organisational learning as a tool to manage change. *Project*, **11**, No. 2, 1998b, 19–20.

KOLB, D. A. and FRY, R. Towards an applied theory of experiential learning. COOPER (ed.). *Theories of Group Processes*, Wiley, New York, 1975, pp 33–58.

KOLB, D. *Learning Style Inventory*, McBer & Co., Boston, 1976.

MAILLAT, D. and KÉBIR, L. Learning region, milieu innovateur et apprentissages collectifs. *Le paradigme de milieu innovateur dans l'économie spatiale contemporaine*. Atti del colloquio GREMI, Parigi, 1998.

MORGAN, K. The learning region; institutions, innovation and regional renewal. *Regional Studies*, No. 31.5, 1997, 491–503.

NONAKA, I. The knowledge creating company. STARKEY, L. (ed.). *How Organisations Learn*. International Thomson Business Press, 1996.

PIETER VAN DIJK, M. and SANDEE, H. (eds). *Innovation and Small Enterprises in the Third World, New Horizons in the Economics of Innovation*. Edward Elgar, 2002.

REVANS, R. *The ABC of Action Learning*, Chartwell, Bratt, Bromley, Kent, 1983.

SAAD, M. Development through technology transfer – creating new organisational and cultural understanding. *Intellect*, 2000.

TIDD, J., BESSANT, J. and PAVITT, K. *Managing Innovation*. 2nd edn, John Wiley & Sons, Chichester, 2001.

TROTT, P. *Innovation Management and New Product Development*. 2nd edn. Financial Times – Prentice Hall, 2002.

Index

Note: Figures, Tables and Case Studies are indicated by *italic page numbers*, footnotes by suffix 'n' (e.g. '19n[2]' means 'footnote [2] on page 19')

Printed in the United Kingdom
by Lightning Source UK Ltd.
9730300002B/45-300